How to ~~Write~~ Edit Your Scientific Article

How to ~~Write~~ Edit Your Scientific Article

Stacey Smith?

The University of Ottawa, Canada

World Scientific

NEW JERSEY · LONDON · SINGAPORE · BEIJING · SHANGHAI · HONG KONG · TAIPEI · CHENNAI · TOKYO

Published by

World Scientific Publishing Co. Pte. Ltd.

5 Toh Tuck Link, Singapore 596224

USA office: 27 Warren Street, Suite 401-402, Hackensack, NJ 07601

UK office: 57 Shelton Street, Covent Garden, London WC2H 9HE

Library of Congress Control Number: 2023014995

British Library Cataloguing-in-Publication Data
A catalogue record for this book is available from the British Library.

HOW TO ~~WRITE~~ ∧EDIT YOUR SCIENTIFIC ARTICLE

ISBN 978-981-124-582-4 (hardcover)
ISBN 978-981-124-684-5 (paperback)
ISBN 978-981-124-583-1 (ebook for institutions)
ISBN 978-981-124-584-8 (ebook for individuals)

For any available supplementary material, please visit
https://www.worldscientific.com/worldscibooks/10.1142/12516#t=suppl

Printed in Singapore

*Dedicated to Lindi Wahl
for being an excellent writer,
an excellent postdoc advisor
and for asking the best question ever*

Preface

Why Can't Scientists Write?

Why are so many academic articles so badly written? Particularly in the sciences, it's shocking just how unreadable most of the output actually is. Articles are drafted by people who excel at the rigour of science, have great logical thinking skills and careful methods... and yet these articles are too often flabby, meandering and awkward, largely because these brilliant people apply precisely none of their scientific skills to the issue of writing.

Publications are the currency of academia. And yet most academic articles, even those by native English speakers, are appallingly written. One of the core problems is that most scientists hate writing and put only the bare minimum of effort into it. Indeed, many people in the sciences — especially mathematics — are never actually taught how to write. More specifically, they are not taught how to edit, redraft and revise their material so that the presentation is optimal for the reader. As a result, academic articles too often read like a first draft, with little understanding that all writing is editing.

Academic writing is a skill like any other that can be broken down into stages. The same principles of logic, thoughtfulness and care that serve scientists so well in their research can be applied to the process of writing. Just as nobody should publish the results of a single experiment, no one should be publishing their first draft... and yet, that's how the vast majority of published scientific articles come across.

This book is based on a workshop I've given in multiple countries (many of them with an audience for whom English was not their first language), and it received enormous acclaim every time. Indeed, the first

time it was presented, in India, the audience started texting their friends
to come, growing the final numbers by more than five times the initial
participants. The second time I gave it, in the Philippines, the audience
numbered over 200. Audience members emailed me the next day to say
that because of this workshop, they completed a stalled project in just
a single night. Other students who saw the workshop subsequently have
reported that it was the most useful workshop they'd ever taken. There's
clearly a strong demand for help with academic writing, because it simply
isn't taught.

In 25 years of teaching at the university level, I've seen science students
nearly panic at the thought of having to write an essay or even just a
report. Students are terrified of writing up their results into a thesis,
while my colleagues — university-trained professors — quake in fear at the
thought of writing a grant application. I've encountered many professional-
development opportunities throughout my career, but writing has barely
registered as a skill that anyone should be taught, despite the obvious need
for it. However, the problem isn't the writing, per se.

It's actually the editing.

Specifically, it almost never happens — or it happens in a kind of
patchwork way based on supervisors' comments or reviewers' instructions
or whatnot. What we as scientists don't learn is how to self-edit: how can
we become our own collaborator and critic? How can we learn to look at our
work dispassionately and find all the holes that need fixing before anyone
else does? How can we take this ineffable and haphazard process that we
call 'writing' and make it algorithmic?

This book is a step-by-step guide to the self-editing process. We'll go
through the detailed process of all the stages, from first drafts to writing
abstracts to revision to responding to reviewers, illustrated with multiple
versions of worked examples... and also examples of what not to do.

The examples are largely drawn from mathematical modelling papers,
because that's my field, but the principles apply much more broadly than
just mathematics papers. Note that I've kept typos and American or British
spelling intact for these examples, because it's important to see the raw
versions.

I should note that I have largely kept the examples of what not to do
anonymous (except where the person who shouldn't have been doing it was
me!). I want to protect the identity of those whose work I'm quoting here,
because I don't want to point fingers. Indeed, I could have plucked these
examples from almost anywhere, because good writing is so rare. Most of

the work was never published, for self-evident reasons. But a few were and are credited as such. However, I want to thank those who did inadvertently contribute, because these are valuable teaching moments, and I appreciate the attempts nonetheless.

What this book is not

This guide isn't about how to come up with ideas or solve equations, although those are of course part of the process. It's not even about how to write a first draft, although there's some of that. And what I've outlined here is just one approach, albeit a successful one. There are others, and none of these rules are sacrosanct... but it's a truism that you should know the rules before you break the rules.

This isn't the first guide to writing, nor will it be the last. Other excellent works include:

- *Science Research Writing for Non-Native Speakers of English [Glasman-Deal (2020)]*
- *The Grant Writer's Handbook: How to Write a Research Proposal and Succeed [Crawley (2015)]*
- *Planning Your Research and How to Write It [Nather (2015)]*
- *Scientific Writing: A Reader and Writer's Guide [Lebrun (2007)]*
- *The Grant Writing and Crowdfunding Guide to Young Investigators in Science [Lebrun and Lebrun (2017)]*
- *Tips and Tools: A Guide to Creative Case Writing [Joshi (2018)]*
- *The 21st Century Guide to Writing Articles in the Biomedical Sciences [Diskin (2018)]*

These books have their primary focus on writing in the first place. What sets *How to ~~Write~~ Edit Your Scientific Paper* apart is the focus on editing, particularly self-editing. Most people who've written a thesis can hammer out words. What is often not taught well (or at all) is the ability to critically tear up your own work and reassemble it. That's what we'll learn here.

Along with examples of what not to do, you'll also find a gradual assembly of a single academic paper, on respiratory syncytial virus. Here, I've been very transparent about presenting the various drafts along the way and the steps taken to create the article at every stage. Most people learn from examples, as I certainly do, so I think showing the nitty-gritty of how an article goes from a single idea to a published paper is very helpful.

A note about mathematics

Being a biomathematician, my field is infectious disease modelling, so a lot of the examples here are drawn from that area (although not all). However, this book isn't just aimed at mathematicians, nor is it even necessarily aimed at scientists. I think everyone in academia could stand to improve their writing (I for one am on a constant journey of self-improvement), so a lot of the tips here are widely applicable.

That said, if mathematics isn't your thing, then don't sweat the equations. There will be some (sorry math-phobes!), but it's the words surrounding them and the ideas they encapsulate that are important here. So if you get bamboozled by the Greek letters, just skim over those parts and take in the ideas, because writing and editing skills are independent of subject area.

About the Author

Usually these 'about the author' sections are a bit of a formality, but in this case, I think it really matters. So here are my writing credentials. I have over 25 books to my name, some academic, but most are not. My pop-culture books are written to simultaneously appeal to the hardcore nerd and the curious newbie, ensuring that both have a good time (and with plenty of jokes to keep things lively).

I've written more than a hundred academic articles, but each one has been crafted and edited with care and attention — and a couple even went viral, becoming media sensations. In addition to being a writer and editor, I also work for my university's academic press, approving other people's books that have merit and guiding them to becoming the best versions that they can be. In my spare time, I write personal memoirs for fun (and just for myself) and edit review websites. In short, writing and (particularly) editing are huge passions of mine.

I'm a full professor at the University of Ottawa in Canada, but I didn't get there the usual way. I started as a working-class kid who was the first in her family to go to university. I had no academic family background, but I had some natural math ability, and I taught myself how to write. That made a huge difference, especially as time went on, because the ability to write a thesis and later academic papers, grants, et cetera, are enormously valuable skills. I rose through the ranks largely on the back of my ability to write, which I freely admit is an odd path for a mathematician, but these days that's the vast majority of what I do in my job day to day. And because those skills were all self-taught, I see no reason why anyone else shouldn't learn them.

You might notice that my name changes between earlier published works and this one. That's because I transitioned from male to female, but I don't shy away from my old name. I don't use it currently, but it's fine to reference it retroactively (e.g., when citing an older work). Academia isn't great about retroactively updating name changes, so I decided to embrace it and keep my old name in circulation with regards to historical documents.[1]

My books have won awards, and I've also won more general awards for my ability to reach beyond academia and connect with the general public. Sometimes this was through pop culture, such as my work on modelling a zombie invasion. Other times it was through media appearances, where I break down complex topics for the lay-person.

And while I'm a decent writer, I'm an *excellent* editor. I've been editing a pop-culture review site[2] — where I turn people's ordinary prose into something workable — every day for several decades. After 10,000 hours of this, you start to get good at it. The key to my writing success is this: as a writer, I know that whatever I write, some amazing editor will come along later and make it sparkle. That person just so happens to be me, with a different hat, but writing and editing are very different skills... and you can absolutely learn both.

I would love nothing more than for academic articles to read smoothly and pleasantly, with the prose not getting in the way of the content. If this book can contribute to making that happen, I'll be a happy academic.

---

[1] Not every trans person does this, so don't generalise from my example. Just ask if in doubt. We don't mind!

[2] The Doctor Who Ratings Guide, at pagefillers.com/dwrg.

Acknowledgements

I'm indebted to Anthony Wilson for his careful beta reading. I talk a lot in here about having extra eyes on the manuscript, and Anthony has been my extra eyes for many, many years. Anthony is a musician and a teacher, but the fact that I feel comfortable sending a book with equations in it to him speaks to the power of having an intelligent non-expert look through the words. I'm a decent writer in part because I trust my beta readers, and Anthony has been my secret weapon for years. For parts of this book, I've brought Anthony's editorial comments in explicitly, because seeing how this very book got revised and improved is part of the process.

Thanks also to Rochelle Kronzek, Liu Nijia and the team at World Scientific for setting this in motion. Thanks to Laura, Mélan, Red, Philémon and Kristen for moral support. I'd also like to thank Daniel Changer, Brenna Frazer, Rachel Batty, Eric Pelot, Kerstin Puschke, Jennel Recoskie, Gina Rosich, Alina Barnett, Ruth Gtom, Mélodie Courval, Elissa Schwartz, Graeme Burk, Arnold Blumberg, Jason A. Miller, Matthew Palmer, Jez Cartner, Miriam David, Catherine David, Katie Moon, Kate Small, Zoé Tulip, Meryki Vagabond, Beck Shepherd, Vince Ringstad, Chris Casimiro, Brittan Fell, Carley Rogers, Alison Kealey, Tara Gallimore, Elliot Chapple, Frankie Calder and Cathy Petőcz for friendship and discussions along the way. Thanks to Jon Blum, fLorance Cotel and the members of the Sweeney Hotel writing group for work-in-progress feedback and to Gabriel(le) DeRooy for turning the final draft into the penultimate one.

Write on!

Contents

List of Figures

Chapter 1

The First Draft: This Is Not Your Manuscript

1.1 Getting started

1.1.1 *Initial steps*

Let's be clear from the outset: Michelangelo did not paint the Sistine Chapel in one go. Da Vinci didn't submit his first sketch of the Mona Lisa for posterity. Alexandros of Antioch did not carve the Venus de Milo perfectly the first time. Likewise, your first draft is not something you should publish.

Almost no words in the first draft should make it through to the final version intact. If they do, you haven't done your job. The first draft needs to be revisited, sifted, sorted, chopped into pieces, re-ordered and rewritten. But — and this is hugely important — don't do it as you write; do that after you're done writing it. We'll examine that editing process in later chapters. So what is the first draft actually for, then?

The first draft is where you find your ideas. Concepts that might be floating around in the aether of your brain need a vestibule to be placed into. Only by capturing them and taming them can your message then be stress-tested for logical consistency, weak spots, missing data and general readability.

Where to start? I think the best place is to write down whatever inspired you to tackle this problem in the first place. It doesn't have to be fancy; we'll find the fancy later on. It doesn't even have to be original (though it helps if it is). It just needs to have some forward momentum.

As promised, we're going to work through an example of a paper in this book, detailing all the drafts and changes aong the way. (We'll also see some examples of what not to do.) Here's the seed for the main example.

WORKED EXAMPLE: *'What if we had a vaccine for Respiratory Syncytial Virus?'*

1

That's it. That was my entire idea at the outset. RSV is a childhood disease that my collaborator was interested in (full disclosure: I'd never heard of it before that). There was no vaccine, but several were in development, according to my collaborator Geoff Mercer, who was more of a biologist than I was. So a natural avenue was to use mathematical modelling to investigate what the implications might be once one arrives.

I didn't know if this had been done before (that's what the literature search will be for). I didn't know if anything would come of it (maybe the vaccine would make no difference). I didn't know if there was going to be anything unique and insightful (that's what the analysis is for). But it gave me a place to start — and to keep going.

(RE)WORKED EXAMPLE: *'If babies need multiple vaccines, could we apply impulsive differential equations?'*

My collaborator mentioned that babies or young children might need multiple vaccinations, so I had an idea to apply impulsive differential equations to the problem. Impulsive differential equations are a technique I studied for my Ph.D., and they make a good fit for talking about repeated actions. I'd done this with other vaccines before, so I knew this idea had potential.

The basic idea is to interrupt continuous processes with sudden shocks or jumps. For example, if you're taking a drug, then your body will metabolise the drug in an approximately exponential decay. If you pop another pill, then the amount of drug will increase sharply. It only takes a short time for the pill to spread throughout the body, so we could approximate this time by an instant jump in drug levels. (See Figure 1.1.) Regular differential equations describe the continuous decay between the jumps, while the impulses tell us how the jump happens (and the jumps don't have to happen at regular intervals). Impulsive differential equations combine the two, forming a semi-continuous system.

BAD EXAMPLE : *'Modeling and dynamics of the novel corona virus disease.'*

This was the title of a manuscript I was sent in October 2020 by some international colleagues. By that point, the COVID-19 epidemic had been underway for more than seven months. Indeed, it had advanced so much that we'd stopped even calling it 'novel coronavirus' by this point. The problem with this example is that there's no hook. There's nothing that tells us what this article is about, other than a very generic investigation

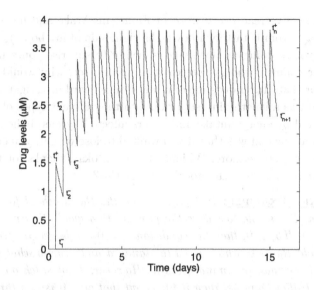

Fig. 1.1. Impulsive differential equations describe how sudden shocks or jumps can be inserted into an otherwise continuous process, such as taking drugs at regular or irregular times.

(of a disease that everyone was investigating). Why should we care? What is this article actually going to be about? And, most importantly, what separates this from all the other investigations out there?

For sure, you might find an original angle during the investigation phase. That's part of what it's for. But your one-line idea still needs legs, and this had none. I simply couldn't tell where this is supposed to go or why anyone should care.

1.1.2 *Find the forward momentum*

Once you have your initial idea, the next step is to move forward. People in Hollywood are always worried about someone stealing their ideas, without realising that an idea is just a small fraction of the finished product. It is absolutely what you do with your idea that matters, not how big it is. So what do you do? Answer: something totally random.

I'm quite serious. You need some time to 'play'. Doodle. Mess around. What you're actually doing, of course, is applying all those excellent background skills you have from your training. In my case, it's mathematical analysis. But it might be intersectional theory. Or data gathering. Or whatever the hell it is that English majors do with the works of James Joyce.

(I'm kidding. I've read every word of Finnegans Wake and have the first and last page from it tattooed down the right side of my body.)

What you're doing at this stage is trying to surprise yourself. If your final paper confirms things you knew at the outset, why would we need research for that? It reminds me of a talk I once heard investigating weight loss. There was a lengthy investigation into the different types of fat cells, some affected by energy intake and others affected by metabolic rate. The conclusion at the end was that if you wanted to lose weight, you either had to eat less or exercise more. Which... I mean... okay, thanks for that, but why did we need a research paper to tell us this?

WORKED EXAMPLE: *In trying to find the R_0 threshold for potential eradication, I was able to reduce the problem to a quadratic equation $\lambda^2 + b_1\lambda + c_1 = 0$. If $b_1 > 0$, then the condition $c_1 = 0$ can be rearranged to form the threshold $R_0 = 1$, which is quite standard and exactly what we want. (This is all normal, preliminary stuff.) However, I got stuck on the issue of $b_1 < 0$. If that happens, then it turns out that $c_1 = 0$ isn't a threshold at all, and hence there is no R_0. Indeed, it's then possible that a disease could persist with 'R_0'< 1, which is counter to every expectation and incredibly misleading, especially because you would find this same 'R_0' value using other techniques that would miss the condition $b_1 > 0$. This had me very puzzled for a while, but it also opened up a huge opportunity to explore. This wasn't where I thought the research was going to go (I hit this hurdle during the earliest stages of exploration!), but it sent me spinning in a whole new direction.*

More specifically, you're searching for the research questions. You want to pose some questions that will eventually get answered. Of course, if you knew the answers from the outset, it wouldn't be research, so the initial questions don't have to be the ones you'll end up with. But it's good to have some to start with, especially if there are things you don't know. How many research questions do you want? Three is probably the best number.

WORKED EXAMPLE:

1. *Can a vaccine make a difference?*
2. *Can we get any insights from impulsive differential equations?*
3. *What happens if $b_1 < 0$?!?*

The second question was my original idea. The third is the one I was super pumped to investigate, because if $b_1 < 0$, then the disease could persist when it shouldn't, and I had a whole new and exciting avenue to pursue. But the first question was important too. Indeed, one of the reviewers essentially made this point later on, because he was clearly a biologist, who was much more interested in whether the vaccine worked or not than quirky math problems that I cared more about. If your paper is likely to have multiple audiences, make sure you have something for all of them. Above all, make sure your questions are crisp.

If you're not sure about this, run them by colleagues in your field and also friends or relatives who aren't. Even if your friends and relatives aren't experts, it's incredibly useful to develop the skill of being able to break down your research into terms a lay person can understand. Ask for honest feedback, and accept it like a grown-up. Don't argue with them, simply thank them... and then go back to the drawing board and come up with better questions. I did this when I was a student, and a lot of my early work was utterly excoriated by some of my friends, who didn't understand math or what the point of it was. I want to stress that this was hugely beneficial to me, because it forced me to think outside my own subdiscipline.

BAD EXAMPLE : *'We find the threshold quantity (Basic Reproductive Number) with the help of the Next Generation Matrix method. We then discuss local as well as the global properties of the proposed model. For the local and global dynamics, we use the linear stability analysis and Lyapnov function theory. Moreover we fit the value of model parameters to the real data as reported by Khyber pukhtunkhwa Pakistan from April 13 to June 9, 2020. For the parameters estimation we use the method of Ordinary Least Square (OLS). Furthermore the detail local and global sensitivity analysis is performed to find the most sensitive parameters and their relative impact of the disease transmission by calculating the sensitivity indices, PRCC and p value of every parameter. Lastly a large scale numerical simulation will be performed to predict the future forecast of the model.'*

This is from the same collaborators who proposed the novel coronavirus idea. This description was their plan, but it's flawed in several places. (I'm not talking about the poorly constructed sentences, which we'll get to in Chapter 5.) The most obvious problem is that there's nothing new here. Every single sentence is the kind of standard thing you'd see in every disease modelling paper. It reads like a shopping list drawn from a thesaurus. And, in contrast with the previous example, there are no questions being asked. They're telling us what they're going to do, without having done it, which leaves no room for exploration or wonder, no place for surprises.

A separate problem is that I don't see what's innovative here or where this would go that would give us any insight that we didn't already possess. Indeed, the only interesting thing about this to me was that it was applying data from Pakistan. But everything else was basically a bit boring, and that's the kiss of death for the reader. I'm not saying you need to throw in a circus act. But your paper doesn't have to be the height of sombre dullness either. This is research; it's not the shipping forecast.

While you're doodling away for original ideas, you're also doing that other thing you're trained for: reading the literature. Search engines are great and can hyper-focus your efforts, but don't forget to actually read. Far too many people attempt to write without reading, and you can tell. It's important to situate your work in the context of what's come before. You'll probably find open questions this way, which may spark ideas.

I like to think of academic research as building a wall. Every time we do research, we add a brick to the wall. So it's crucial to know where your brick fits. We'll come back to how to write the Introduction later, but this literature search is part of your data gathering.

1.1.3 *Methods*

The next thing to do is to figure out the Methods section. In the case of disease modelling, that involves creating the model itself. But in other fields it might involve figuring out what data you can realistically gather or what theories you're going to draw upon. As always, these don't have to survive the process, but you need something to start with.

WORKED EXAMPLE: In my case, I extended an existing non-vaccination model (from Weber *et al.* (2001)) to add vaccination. The following was my initial outline of the model:

The basic model with vaccination is

$$S' = \mu(1 - \epsilon p) - \mu S - \beta(t)S(I + I_V) + \gamma R + \omega V$$

$$I' = \beta(t)S(I + I_V) - \nu I - \mu I + \omega I_V$$

$$R' = \nu I - \mu R - \gamma R + \omega R_V$$

$$V' = \epsilon p \mu - \mu V - \beta_V(t)V(I + I_V) + \gamma_V R_V - \omega V$$

$$I_V' = \beta_V(t)V(I + I_V) - \nu_V I_V - \mu I_V - \omega I_V$$

$$R_V' = \nu_V I_V - \mu R_V - \gamma_V R_V - \omega R_V$$

with $\beta(t) = b_0(1 + \bar{b}\cos(2\pi t + \phi))$ and $\beta_V(t) = (1 - \alpha)\beta(t)$, for $0 \leq \alpha \leq 1$. *(We may relax the lower bound on α later.)*

The first three equations come from Weber *et al.*, with the last three (and a small tweak to the first equation) the new material. They're the obvious extension, based on my years of experience with vaccine models. The form of β comes from Weber *et al.*, while the caveat about α is because I already knew this was what would be needed to get $b_1 < 0$.

(If math isn't your thing, don't sweat the details. The important thing here is that I've established a premise from which the results will inexorably proceed.)

The next thing to do here is the standard analysis: stuff that needs to be done but won't set the world on fire. There are basic things that every paper in your field needs, so you need to do them too.

WARNING: *Doing this does not make your results worth publishing!*

I've seen way too many manuscripts that simply replicate stuff any advanced undergraduate or grad student could do and attempt to pass these off as publishable. (See the coronavirus bad example above.) They're necessary but not sufficient.

WORKED EXAMPLE: *I found the disease-free equilibrium, made a simplifying assumption about β (making it constant for the purposes of the linear stability analysis), created the Jacobian matrix and found R_0.*

The final thing to do in the first draft is to decide what the pictures are going to look like. You definitely want at least a rough idea of this early on. Some subfields aren't heavy on images, but I think this is generally a mistake. People learn in all sorts of ways, and visual aids are immensely helpful.

Let's be clear: most people in this world aren't going to read your paper. Indeed, most people who glance at your paper aren't actually going to read it. Generally, people who read academic papers can be divided into groups, in descending order of population. See Figure 1.2.

This means the figure captions — and hence the figures themselves — have to tell the story of the paper, independent of the actual text. I'm way less of a visual person than a words one, but I've had reviewers of my books tell me that I absolutely need to spice up my text with images, and they were right. Is this easy? Not always. Is it necessary? Yes.

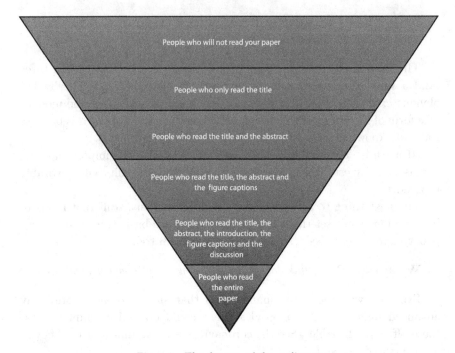

Fig. 1.2. The descent of the audience.

WORKED EXAMPLE: Here were my four initial figures. *Figure 1.3(a) shows the total number of infected individuals when $p = 0$; i.e., when there is no vaccination. Figure 1.3(b) shows the total number of infected individuals (whether vaccinated or not) when $p = 1$; i.e., when everyone is vaccinated. The two figures below them show the total number of individuals in each class after 150 years.*

The four panels in Figure 1.3 are pretty terrible figures. The first one doesn't even have labels on the axes. The last two have the data in a text legend (how gauche!). The timescale is way too long. It's hard to tell what the take-home message here is unless you look REALLY close at the y-axis and discover there's a slight drop from 7% to 6%. That's it. That's the entire visual story at this point.

But that's not the point. The point is that a story is already emerging: the vaccine will make things better. The final size for I and I_V in

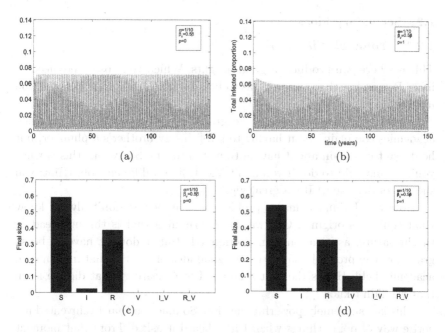

Fig. 1.3. (a) Total infected, no vaccine. (b) Total infected, complete vaccination. (c) Final size, no vaccine. (d) Final size, complete vaccination.

Figure 1.3(d) is lower than the final size of I in Figure 1.3(c). That's something. It's not enough, but it's something.

1.1.4 *What the first draft does not include*

Here is a helpful list of things that do not belong in the first draft:

- Abstract (always write this at the very end).
- Introduction (you can't introduce a topic you haven't done yet).
- Discussion (you can't discuss the implications of a topic you haven't done yet).
- References (you can — and should — compile them as you go, but they don't need to be in the draft).

As always, you can break any of these rules... once you know the rules. (And I mean *really* know them, having implemented them many times first.) The point of the first draft is to get your ideas down and let them take you in a surprising direction. It's not to write the paper. That comes later.

Much, much later.

1.2   Secondary ideas

1.2.1   *Your job (literally)*

This is where you produce original insights. Which you absolutely need to have! Remember that research has to be original. This is your job, and you have to do it.

I come from a working-class background, so it always amuses me when academics complain about having to publish. My brother's a plumber, but he doesn't complain about having to install toilets. Sorry, but this is what you're being paid to do. If you don't like it, find a different job. (Hint: you should totally like it! It's a great gig.)

However, I think a major problem is that people don't always know what counts as original. One way of determining that is through reading the literature, as mentioned in Section 1.1. But it doesn't have to be so granular. You probably have a pretty good idea of what's what in your own academic field. Here's the best sentence I ever heard: 'What did we learn that we didn't know?'[1]

This is a supremely powerful question. So much so that I reinvented my entire way of doing things when I first heard it asked. From that moment on, that question was at the forefront of everything I did, because this is absolutely what we should be doing when we do research. If everything you do is something we already knew, then what's the point of research?

Where the first section was about getting through the standard stuff, this is where your doodling pays off. It might go in a direction only vaguely related to the first part, which is perfectly fine. Maybe you solve a long-outstanding problem. Maybe you have insights that no one else has had. But you need *something*. Experience should tell you when you have it, but if you don't know how to judge that, I suggest running it by a more senior colleague in your field.

Oh, and if you don't have something original, then keep doodling. Worst-case scenario, the project gets abandoned at this early stage. That's way better than being forced to abandon it after you've put in all the work to write and revise and design figures. Or, worse, publishing substandard work and gaining a reputation as a weak researcher. Your good name and your reputation are your foundations as a researcher, so don't build weak ones.

---

[1]I can't claim credit for this. It's from my brilliant postdoc advisor Lindi Wahl, to whom this book is dedicated.

What you're doing here is finding your research question. Or the first one, at least. (Remember, you'll need three.) This question should be something that's interesting, original and surprising — to you. If you can't surprise yourself, your chances of surprising anyone else are poor. What jumps out at you as an amazing insight, one that has you wild-eyed and alert? You'll know it when you find it. That's your research question.

WORKED EXAMPLE: *In trying to tackle my $b_1 < 0$ problem, the first issue was to see if it even actually occurred. I picked a few parameter values at random and plugged them into the model to see if I could make it happen.*

In doing so, I realised that one parameter (ν_V, the recovery rate for vaccinated kids) was having more effect than others. I chose some pretty wild parameter values to try this, but at this stage I was more interested in the possibility of it occurring than I was the likelihood that it would in reality.

I also noticed that if $\nu_V = \nu$ or if $\nu_V = \infty$, then $b_1 > 0$. Why these values? Because they're the two most extreme ones. No vaccine is going to be released that causes kids to be sicker than the actual disease, so that means ν_V has to be greater than ν, the recovery rate without vaccination. The other extreme is that everyone recovers instantaneously, which corresponds to $\mu_V = \infty$ (because the rate is inversely proportional to the time spent in that class).

It only took a few funky ν_V values before I realised that it was indeed possible to make $b_1 < 0$. So now my research question wasn't just theoretical! I ran this by my more senior co-author, Geoff, who immediately asked if it was a backward bifurcation. It wasn't, but the fact that he thought that meant I was going to have to point that out upfront, because most readers would probably think the same thing initially.

A bifurcation is something that happens when the system undergoes a fundamental change. When R_0 moves from below 1 to above 1, the number of equilibria increase, and the stability of the disease-free equilibrium changes. A forward bifurcation occurs when the system shifts from one equilibrium to more than one (usually two) as R_0 increases. Likewise, if R_0 could be reduced below 1, then there would only be one equilibrium (the good one) and the disease would be eliminated, because of that equilibrium's stability. But sometimes things go haywire, and reducing R_0 below 1 doesn't correspond to only a single equilibrium existing. We call these backward bifurcations, and they cause trouble because the disease can persist for $R_0 < 1$, assuming the initial conditions are high enough. (If the initial conditions are sufficiently *small, then the same properties hold as the forward equilibrium.) For the RSV model, if $b_1 < 0$, then the disease could persist regardless of initial conditions, and that's a serious problem for eradication.*

What did I learn that we didn't know? It's possible to destabilise the equilibrium in a new way. Cool!

I want to stress that there's no algorithm for this part. Unlike the rest of this book, which can be used as a roadmap, finding your research question isn't a standard process. This is where you bring your experience and insights to bear. Think outside the box. Combine unlikely concepts. Go in wild directions. You've been trained for this (that's what a Ph.D. is for), so use it. Impress us.

BAD EXAMPLE : *There are many examples of the same old, same old papers in the literature. Far too many papers get published that lack insights and just seem to be spinning their wheels. You've probably read a million of them. But I'm going to turn inwards for this one and include one of my own papers as a bad example [Aggarwala and Smith? (2009)]. Feel free to check it out and cringe. But let me illustrate the problem by summarising the section titles in the paper:*

(1) *Introduction*
(2) *Parameters*
(3) *Positivity of solutions*
(4) *Boundedness of solutions*
(5) *Equilibria*
(6) *Stability of equilibria*
(7) *Viral blips [this is a very short section, only a single paragraph]*
(8) *Stability of the endemic equilibrium*
(9) *Varying parameters*
(10) *Conclusion*

If you know anything about differential equations, you'll see that there's absolutely nothing here that isn't very standard. We did the basic investigation of the model (positivity and boundedness), found and analysed equilibria (yawn) and then ran some numerical simulations (boring). These things are all necessary but in no way sufficient. What did we learn that we didn't know? Absolutely nothing. This was published before I heard that question, and you can tell. (If you don't know anything about differential equations, then a good analogy would be a research paper that retells the plot of Macbeth, whilst noting that it might be about the concept of kingship.)

Sadly, this isn't unusual. Far too many papers do this, when it's pretty much the exact opposite of what original research is supposed to be about. Don't make my mistakes.

WARNING: *What did we learn that we didn't know? Make sure you actually answer this question.*

1.2.2 Refining the secondary analysis

Okay, so you've had your breakthrough! You've found an original insight, and it's exciting and new. Time to publish the paper, right?

Not so fast.

Like that old Hollywood problem, this is the beginning of the process, not the end. You absolutely need that original insight. But you also need to develop it. Specifically, what are the implications?

It's very likely that your original insight will lead to further insights. I said before that you probably need three research questions for a paper, so this is a chance to maybe pick up another one. It's crucial to know what happens because of your insight. What are the consequences? What does it all mean?

WORKED EXAMPLE: *Because $b_1 > 0$ at the two endpoints, if $b_1 < 0$ is going to happen, there has to be a critical point. Under what conditions would a critical point exist? Can we find it?*

A critical point occurs when the derivatives are zero, and it's a place of stasis, where nothing can change. If the critical point is a local maximum, it's the highest point in its neighbourhood, like a mountain peak. If the critical point is a local minimum, then it's the valley.

It turns out that I could find the critical point... but then I hit another problem. I hadn't checked to see if the critical point was actually a local minimum. It might have been a maximum, which wouldn't have been helpful. So let's add in some more questions.

(RE)WORKED EXAMPLE: *Is it definitely a local minimum? Can we prove that?*

You'll notice what I'm doing here: trying to plug the holes. That's going to be a theme of this book, not just in the analysis section but in the writing section as well. It's crucial to identify weaknesses and bolster them. The same skills that teach you to apply logic, rigour and critical analysis to a

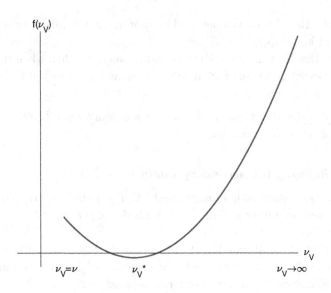

Fig. 1.4. A conceptual diagram for how μ_V could make $b_1 < 0$.

research problem can be harnessed when writing. So hold onto this thought, because we'll definitely be coming back to it.

Once again, a figure is helpful, even if it's just a conceptual one. In the upside-down pyramid of Figure 1.2, the first layer where people actually turn the pages is the one that has the figure captions. What are these skim-readers looking for here? Answer: a narrative.

WORKED EXAMPLE: *Figure 1.4 is a sketch I made of what I thought — hoped, really — was happening. I had no idea if it was even possible, but it outlined what I was looking for, and that helped organise my thoughts. It's a purely conceptual diagram, because I had no idea if the minimum even crossed the x-axis, but it helped me tell the story I was uncovering.*

I might not have included this figure in the published paper, because it was really just a tool for organising my thoughts. In the end, I decided to include it, because I thought it would help the reader as much as it helped me.

BAD EXAMPLE : *Check out Figure 1.5. I found this figure in a linguistics paper that conducted a statistical analysis of different speakers in New York department stores [Guy (2018)]. I'm not qualified to talk about the linguistics component. But I can say that this is probably the single worst figure I've ever seen in a published paper.*

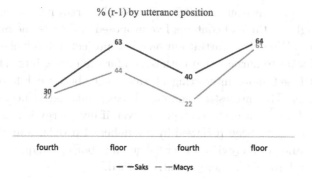

Fig. 1.5. The worst published figure I've ever seen.

What on earth is Figure 1.5 trying to convey? There's no y-axis label. The x-axis labels are laughable, and I can't even begin to decipher the progression. There's absolutely no reason to link the numbers with lines in the first place. The legends don't even match (they're solid lines, not dashed).

Your paper needs to tell a story. Like any story, it needs to have a beginning, a middle and an end. It needs to intrigue, inform and entertain. It needs to have twists and unexpected developments. And by the end, it needs to feel satisfying.

Perhaps the greatest superpower humans have is our ability to tell stories. We've been doing this ever since we huddled around fires in caves. We're very, very good at it. So harness this superpower you have. What story do you want to tell?

The rest of this book is about how you're going to tell that story. And the story isn't fully formed yet. But you need this section in order to know what your story is going to be about in the first place.

1.3 Further ideas

Okay, so you have your initial research question and some interesting implications. Your paper still needs more. Ideally, you want another one or two of these. As mentioned earlier, three is a good number of research questions, so that means you need (at least) three big insights per paper.

At this stage, these next steps don't need to be complete, but they do need to be written down. If you have enough expertise and experience, it's fine to leave these and come back to it. If not, write it up now.

Here's a pro-tip from a writer: whenever I start a new book, I always look through the table of contents I've proposed and write the most boring chapter first. When I'm starting out on a new project, I'm full of enthusiasm for it, so I want to harness that enthusiasm for the boring bits. That boring chapter will get more interesting at this stage, because I have a lot of energy for it. The interesting chapters I leave until later, because they'll have enough momentum to keep going, even if my energy is flagging.

My books have been reviewed by a number of people, sometimes quite critically... and yet no reviewer ever flagged my 'boring' chapter as one that was substandard. That energy clearly worked.[2]

WORKED EXAMPLE: *I knew I wanted to add in impulses to the model (this was my original idea, after all). So I simply wrote down the impulsive model. In my original draft, this was the entirety of Section 4:*

$$S' = \mu - \mu S - \beta(t)S(I + I_V) + \gamma R + \omega V \qquad t \neq t_k$$

$$I' = \beta(t)S(I + I_V) - \nu I - \mu I + \omega I_V \qquad t \neq t_k$$

$$R' = \nu I - \mu R - \gamma R + \omega R_V \qquad t \neq t_k$$

$$V' = -\mu V - \beta_V(t)V(I + I_V) + \gamma_V R_V - \omega V \qquad t \neq t_k$$

$$I_V' = \beta_V V(I + I_V) - \nu_V I_V - \mu I_V - \omega I_V \qquad t \neq t_k$$

$$R_V' = \nu_V I_V - \mu R_V - \gamma_V R_V - \omega R_V \qquad t \neq t_k$$

$$\Delta S = -rS \qquad t = t_k$$

$$\Delta V = rS \qquad t = t_k$$

That's it. It's just the model, and nothing else. Why? Because I was on very solid ground with this kind of model (I did my Ph.D. in this topic), so I knew I'd be able to come back later and fill in the analysis.

Once I had the model, it was easy to run some numerical simulations to get the basic results. Note that, once again, these are necessary — but not sufficient. Figure 1.6 shows the baseline outcome of a model with vaccination. There are susceptible, infected and recovered individuals both unvaccinated (in large numbers) and vaccinated (in much smaller numbers). After a few oscillations, things settle down. That's all lovely. It's just not groundbreaking. Yet.

---

[2]See if you can guess from the energy levels which was the first chapter I wrote of this book.

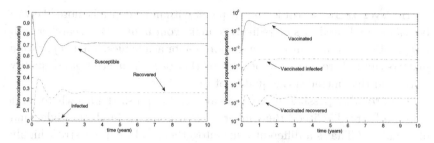

Fig. 1.6. The model with vaccination, showing very standard results.

You'll note that, originally, this is what I thought my article would be. But in the end, it's just a section of the article, and not even the most important one (though it'll go places).

1.3.1 *A word about paper quality*

The other important thing to do is to follow through on the exciting ideas you've found in the previous section. It's not enough to just have a good idea. You need to go down the rabbit hole and find out where it takes you.

Along the way, there will probably be issues that come up that you might need to deal with later. That's okay. The important thing is to flag them so you don't forget them.

What I'm stressing here is depth. Each paper needs academic depth and original thought. You need to have the ideas in the first place, but you also need to take those ideas to the places they want to go. And you have to be able to do this again and again, often in the same paper. Remember: it's your job to come up with ideas. You have lots of them. So don't hoard them, use them in abundance.

Is it better to have one good paper or several lesser papers? There can be a temptation to break a paper into smaller bite-size chunks and publish each of them. This is sometimes referred to as the 'smallest publishable unit'. There are pros and cons to each approach, of course.

Publishing is the currency of academia, but in recent years there's been a trend towards 'more' papers, rather than 'better' papers. Administrators will often be more impressed by a string of papers, regardless of quality, rather than a few good ones in impressive journals. One thing about bean-counters is that they can count. With the pressure to publish, it's understandable why some people will try to publish as many papers as they can, rather than put energy into the quality of those papers.

However, there's a cost to this. A paper with more depth is more likely to be read — and cited. It's more likely to make your name as a researcher with a reputation among your peers and more senior academics. That reputation matters, and there are no shortcuts to this part. You need good quality work, and that includes the quality of the writing.

It's probably possible to get a weak paper published somewhere. Is it going to be read? Cited? Remembered? Is it going to catch the right kind of attention? That's a different issue altogether, and I cannot stress highly enough how much you should put energy into writing strong papers.

WARNING: *Superficial or single results should not be published!*

Macbeth really is about kingship. That doesn't make your observation worth publishing. (Because it's not just about kingship.)

WORKED EXAMPLE: *I did further numerical simulations to see if I could figure out if $b_1 < 0$ was even possible. I trawled through parameter sets (basically by using the ideas from Figure 1.4 to guess some numbers) to find out whether I could see it. Good news: I could. Even better news: I found something entirely unexpected. See Figure 1.7.*

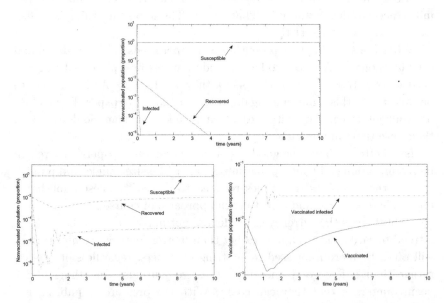

Fig. 1.7. Unexpected infection spikes. Top row: no vaccine. Bottom row: 100% vaccination coverage.

What's so strange about this? It's the case where 100% of the population is vaccinated. I would have thought that any cases would disappear very quickly, but in fact they don't. Even stranger, thanks to the parameters I chose, this is the case when there's no disease without the vaccine... but vaccinating everyone not only causes the infection to emerge but it produces these strange spikes. What on earth is going on? One thing's for sure: we're definitely learning things we didn't know.

A side note about parameters for the biomathematicians reading this. They should absolutely come from the biological literature, not other mathematical modelling papers. You should also provide references for every parameter you use. It's shocking to me how rarely this is done in the mathematical literature. If you want to go beyond the existing values, as I did here, that's perfectly fine... just be ethical about it. State clearly which ones are biologically reasonable (with citations), which ones you've calculated (with details in the main part of the paper or an appendix) and which are extreme values that illustrate a theoretical point.

BAD EXAMPLE : *This example is from a submitted Master's thesis that had — I kid you not — 190 figures. Figure 1.8 was typical. The three figures presented here are near-identical versions with slightly varying R_0 and phase-difference values... so why do we need three of them?*

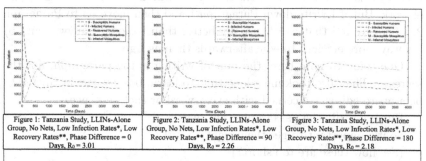

Figure 1: Tanzania Study, LLINs-Alone Group, No Nets, Low Infection Rates*, Low Recovery Rates**, Phase Difference = 0 Days, R_0 = 3.01	Figure 2: Tanzania Study, LLINs-Alone Group, No Nets, Low Infection Rates*, Low Recovery Rates**, Phase Difference = 90 Days, R_0 = 2.26	Figure 3: Tanzania Study, LLINs-Alone Group, No Nets, Low Infection Rates*, Low Recovery Rates**, Phase Difference = 180 Days, R_0 = 2.18

*Low, Middle-level, and high malaria infection rates are based on the minimum tested range value, the value in the Parameter Values Tested in Time-Series Analyses section, and the maximum tested range value respectively for the average malaria infection rates for humans (b_0) and mosquitoes (b_2) from Appendix A related to each specific study that the values were tested

**Low, Middle, and high recovery rates are based on the minimum tested range value, the value in the Parameter Values Tested in Time-Series Analyses section, and the maximum tested range value respectively for the Infected to Susceptible Recovery Rate (h) and the Infected to Recovered/Immune Recovery Rate (α) from Appendix A related to each specific study that the values were tested

Fig. 1.8. Unexciting figures.

However, it's not just the figures themselves. The captions are perfectly descriptive — and entirely off-putting. The footnotes are awkwardly written (the '...were tested' at the end hangs there artlessly). And figure legends — the boxes inside the figures — are almost never a good idea; they make the reader do needless work to process the information.

The rest of this book is about how to present your ideas, not how to come up with them. There are other places for that, including your educational background. But it's important to make sure you have something substantial in the first place. All the great writing in the world can't make up for substandard work.

1.3.2   *How to begin writing your manuscript*

For many of us (myself included), the hardest part about writing is actually starting. Once I have something, momentum can usually carry me forward. But getting words down in the first place can be a big challenge. It took me a while to identify my problem here: it was because I kept wanting those words to be perfect.

A philosophy professor of mine, Vic Dudman, once told me 'You don't need to get Paragraph 1 perfect before you begin Paragraph 2.' Those were life-changing words for me. I still struggle with it, because when I first write something, I always think it sounds corny or half-hearted. That's usually because it is... on account of the fact that I haven't yet done the majority of work. But if I write down Paragraph 1 and then move on to Paragraph 2, I tend to find that a) it's usually better than I thought anyway and b) it's way more efficient to edit afterwards than as I go.

One thing to note here: the unit of writing is the paragraph, not the sentence. That also took me a long time to learn, because my instinct is to be a micro-editor, concerned with every word or each sentence.

But it's really the paragraphs that matter.

So it's important to get them right.

(See how these are too short?)

1.3.3   *The skeleton*

The single best way of getting those paragraphs down is to outline a skeleton in advance. For each paragraph, write a single sentence outlining what that paragraph is about. You're not married to these sentences; in fact, none of these sentences will actually appear in the manuscript itself, because they're just a rough guide for your own purposes.

Once you have the skeleton, you have a global view of what your article is going to look like. From here, you can very easily swap paragraphs around or delete one that isn't necessary or add something that was missing. And since you haven't actually written the paragraphs yet, there's no cost to this. You're not losing precious words when you delete a paragraph. It's incredibly freeing.

Here's the skeleton I wrote for this very section you're reading right now:

WORKED EXAMPLE:

- *Further directions*
- *→doesn't need to be completed*
- *Example: sketch of impulses*
- *Follow the ideas where they lead you*
- *Identify issues that arise*
- *Superficial or single results should not be published*
- *One good paper vs two lesser papers*
- *Bean counters can count*
- *More depth = more likely to be accepted (and read)*
- *Bad example: malaria-spraying abstract*
- *Communication: if you want to change the world, you need to communicate your ideas*
- *If you don't write it down, you didn't do it*
- *More pictures*
- *Eg numerical simulations*
- *Parameters*
- *Bad example: Tanzania figures*
- *What do we learn that we didn't know?*
- *Eg Infection spikes*

You can follow through this section and see that a lot of this is here, but the order got shuffled and other paragraphs got added (including this part!) and some bits got deleted. However, this skeleton was invaluable for giving me a guide when I started to write the section. Without it, I would have been floundering, trying to find the narrative . With it, I simply expanded on each bullet point to make a paragraph. Later revisions involved more expansions and tightening and reworking, but the skeleton remained mostly intact.

If you think writing a manuscript looks daunting, imagine how terrifying starting an entire book is. I use this method for every book I write. It turns the process into something consisting of bite-size chunks and allows me to first map out the shape of the book, then fill in the details. It makes the writing itself more autonomous, which is a godsend. And it saves me from writer's block.

1.3.4   *Overcoming writer's block*

Here's a radical statement: I don't think writer's block exists. Instead, it's something that people run into because they've skipped several steps. If I want to write my novel and haven't planned it out and am simply waiting for the muse to strike, that's a really tough call. But if I'm about to start writing my novel and have an outline of the plot, a skeleton summary of every scene, a pile of background research and detailed character biographies, it's a lot less of an uphill climb.

The same applies to academic writing. Nothing stops you writing a sentence that will outline what you want to say in a paragraph, especially when — and this is crucial — that sentence will not appear in the final document. It won't even appear in the first draft. So what's stopping you writing utter nonsense? Absolutely nothing.

Even better, when you have some expertise in a subject (e.g., from a degree or two in it), then you already know stuff. It's easy to write these things down. They probably won't set the world on fire, but they don't need to. And, crucially, you won't be staring at a blank page.

Once you have that skeleton, fleshing out each sentence into a paragraph isn't very difficult at all. If it is, then just write more of the skeleton. Or do some mathematical analysis.

In my case, I often do some math until I get stuck, then I switch over to some writing. When I'm stuck there, I go back to the math... and having switched gears entirely, I usually find that I'm unstuck. When I get back to the writing again, I'm a lot more fresh. So this way, I'm never actually stuck, because I have a way of fooling my own brain into accomplishing twice as much as I intended to. 10/10 would recommend this method.

1.3.5   *Why bother to communicate well?*

Let's flip this around. Suppose you have great initial ideas that you've followed down the rabbit hole. You have interesting, exciting results. Why should anyone care how they're written down?

The simple truth is that your readers are human beings. We care about these things, even when we don't intend to. Good writing is not strictly necessary, but it's a sign of sophistication and confidence, of the finer things in life.

There's a golden rule of academia, which says that if you didn't write it down, you didn't do it. You can run all the experiments you want, understand all the literary symbolism you like, but if you don't actually create something out of this and write it up for others to read, then you're just whistling in the dark. Isaac Newton attributed his ability to see further than anyone else to his standing on the shoulders of giants. Writing is our way of providing our own shoulders for the next generation to see a little further ahead.

So if you want to change the world, you need to communicate your ideas. And the better you can communicate them, the more likely you are to make those changes stick. I've had utterly brilliant insights that could have changed the world... but those insights were buried inside papers full of math, which weren't read by the people who could actually turn them into policy. It's not just the ideas that matter, it's how to communicate them.

BAD EXAMPLE : *'The aim of this work is to observe the effect of spray insecticides as a means of preventive control against malaria. All this by using the impulsive differential equations, solving by Laplace and Hankel methods. Performing simulations on fixed spray periods via Python software. It appears that for an instant source point at the initial condition and sprays at fixed times, the mosquito population is controlled up to a certain threshold. Demonstrating possible resistance on their part.'*

This was an Abstract from a project that was actually pretty decent once you got into it. But you'd never know it from this atrociously written paragraph. The sentence structure is clipped and awkward. But it's more than just the word choice; the paragraph tells us what's being done, but not why we should care.

Novelist Jim Lewis (author of *The King is Dead*) would tape every page from his book around his house, at ground-floor level. He would assess each page and make improvements, raising the pages higher when they were better. Each page would inch itself up the wall as he made corrections and modifications. Slowly, the entire book would rise. When every page had reached the ceiling, that's when he was ready to publish it.

Your academic paper should be the same. The first draft is merely ground level. What you need to do next is to revise it and refine it until it starts to rise, page by page or section by section. Only when it reaches ceiling level is it ready to be shown to the outside world.

1.4   Stress-testing your basic ideas

All writing is editing. That's a theme we'll be returning to over and over again in this book. We'll get to the micro-level editing later. But for now, we need to think about the macro-level editing. There's no point writing beautiful words if your fundamental ideas are lacking. So we need to refine the basic ideas before we go on.

The scientific method involves throwing lots of questions at a thing and trying to see if it'll survive them. Your ideas are no different. You need to stress-test your ideas in order to make them stronger. (We'll do the same with the writing itself later on.) Hit your ideas with all the counterfactuals and what-ifs. See what happens when they encounter an attack they weren't anticipating. Can they survive it? Will this open up further opportunities?

WORKED EXAMPLE:

(1) *What if the vaccine coverage varies?*
(2) *What happens when there's no vaccine?*
(3) *What if the vaccine lasts forever?*

Here are some spitballing questions I threw at my problem. For the first question, I needed to see what the possible outcomes might be, since there was no fixed vaccination level. For the second question, I needed a baseline case. For the third, I wanted the other extreme. It's highly unlikely these vaccines will last very long at all in reality, but it's also good to know the upper limit of what we might be dealing with.

These cases are illustrated in Figure 1.9. As you can see, there's not a whole lot of difference between the first and second cases, so the varying coverage isn't really telling us anything yet. The third case is a bit more dramatic. Even more so, because the scale on the final axis is only up to 0.5, when the others are 0.7. In a published paper, I would have definitely standardised these scales. But it's also possible to see that even the best vaccine isn't going to eradicate this disease, so that's something we've learned.

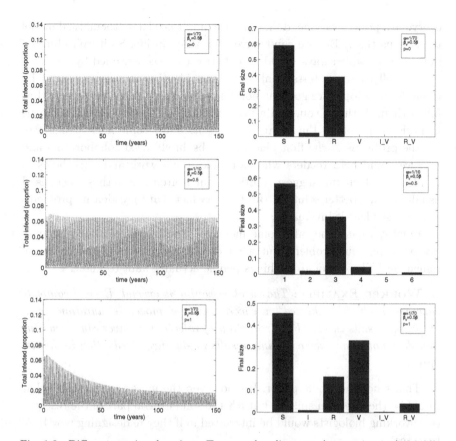

Fig. 1.9. Different vaccine durations. Top row: baseline case (no vaccination). Middle row: intermediate case (vaccine lasts for 10 years, 50% coverage). Bottom row: extreme case (vaccine lasts for 70 years, complete coverage). Note the lack of labels on the right graph of the middle row, because this is such a first draft that I didn't even label properly.

It's also good to identify the key features of your manuscript. If you can't summarise the central ideas in a few sentences, writing more is not going to help. Specifically, what's the central contrast? It's good to write down true facts about a thing, but at some point you need to have more of a story to tell. A story isn't just a litany of facts. One of the best ways to do this is to identify the key contrasts in your work.

WORKED EXAMPLE: *Which is better: a good vaccine with weak coverage or a weak vaccine with good coverage?*

This is an intriguing question, far more so than the earlier example of a lifelong vaccine that everyone gets. The best option would be to give everyone a perfect vaccine, duh. (And we didn't need a mathematical model to show us that!) But we can't have everything in life. So if we're limited in some way, which ways would it be better to be constrained by?

As well as stress tests, make sure you identify any issues... and fix them. Some people are good at noticing problems but not always good at fixing them. If that's you, find a workaround. I'm the opposite: if someone identifies a problem, I'm great at fixing it, but I sometimes struggle to see the problems in the first place. So I absolutely need collaborators and reviewers and beta readers who can look at my work and poke holes in it. Many of them will suggest solutions, but I often ignore those, because I usually know a better solution. But the very fact that they identify problems in the first place is invaluable.

Finally, what's the take-home message? You'll expend a lot of energy investigating your problem, but what do you actually want the reader to remember? These should be things people actually want to know.

WORKED EXAMPLE: *The vaccine waning is crucial. Even if complete coverage could be achieved, a vaccine with a moderate duration (e.g., 10 years) results in very little reduction of infection. Conversely, a vaccine that does not wane over a lifetime results in significant reduction in disease burden.*

That's not the whole paper. It's not even the most exciting part (that would be the infection spikes), but it's the solid central part of the research that working biologists would be interested in if they're designing new RSV vaccines.

BAD EXAMPLE : *'Pine wilt disease (PWD) is a lethal disease caused by a native nematode and destructive in terms of environmental losses and economic cost. Once affected, the trees are destroyed within a few months. Moreover, the asymptomatic carrier cases play a crucial role in the subsequent persistent of pine wilt disease. The present investigation deals with a new mathematical model of pine wilt disease by incorporating the asymptomatic carrier classes. We determine the epidemic threshold of the basic reproduction number $\mathcal{R}_0$ and note that the infection-free equilibrium is global asymptotically stable (GAS) if and only if $\mathcal{R}_0 < 1$, while the endemic equilibrium of the system is GAS for $\mathcal{R}_0 > 1$, i.e., the pine wilt disease persists in the system for $\mathcal{R}_0 > 1$. A thorough sensitivity analysis using the PRCC technique for detecting critical factors that significantly affect the pine wilt disease has been conducted. Concomitantly, the optimal control theory is applied to obtain the key factors which reduce the level of infection* *by using three control measures: deforestation of affected pine trees, the tree injection and the aerial spraying of insecticide on the wounds of pine seedlings made by sawyer beetles. We also investigate analytical condition which ensures the existence of the optimal control solution. Moreover, for a thorough quantitative insight of the model considered, numerical simulations are carried out to show the effectiveness of proposed control measures and provide better suggestions for control interventions for the concern authorities.'*

This draft Abstract was technically accurate, but (a) boring, (b) unreadable and (c) missing its core idea. What's this really about? Three control measures are applied, but we're not told which one is best or what the outcomes are. Or, to put it another way, what did we learn that we didn't know?

(RE)WORKED EXAMPLE: *'Pine wilt disease is a lethal tree disease caused by nematodes carried by pine sawyer beetles. Once affected, the trees are destroyed within a few months, resulting in significant environmental and economic losses. The role of asymptomatic carrier trees in the disease dynamics remains unclear. We developed a mathematical model to investigate the effect of asymptomatic carriers on the long-term outcome of the disease. We performed a stability and sensitivity analysis to identify key parameters and used optimal control to examine several intervention options. Our model shows that, with the application of suitable controls, the disease can be eliminated in the vector population and all tree populations except for asymptomatic carriers. Of the possible controls (tree injection, elimination of infected trees, insecticide spraying), we determined that elimination of infected trees is crucial. However, if the costs of insecticide spraying increase, it can be supplemented (although not replaced entirely) by tree injection, so long as some spraying is still undertaken.'*

It took some digging to discover what this manuscript was about. When I found it, it was actually pretty awesome. There's a clear answer to the question (cutting down infected trees is best), as well as an unexpected implication (when the insecticide spraying costs increase). But the core ideas had been smothered by the impenetrable writing. This is a classic example of poor writing doing the science no favours.

This article went from being a hot mess to being published in a decent journal (Khan *et al.*, 2020), with a respectable number of citations. That's a classic example of the rewriting process saving the manuscript from itself.

1.5   Implications

Here's the next question to throw at your work in progress: do these results make sense?

You should be throwing lots of questions at your work, but this is perhaps the most important one. Academics often deal with the minutiae and the odd things around the fringes, which are nifty and exciting and can yield interested tidbits... but we sometimes forget the big picture. That's not really our fault, because we're rarely taught big-picture thinking, but it's something we need to figure out sooner or later.

It's crucial to think through the potential implications of your work. Because other people certainly will. If you submit something, reviewers are going to treat your ideas seriously, but they will want to see where you're going to take those ideas. If you turn your research into a talk, you'll be hit with questions about what it all means. And if you get something published, you want to make sure you can defend it and have thought through all the places this might go. Because sometimes those places can be embarrassing and awkward — or even downright wrong — and you're much better knowing that in advance and dealing with them than having someone else find out later.

Too many academic papers just publish the next step, without following the logic and seeing where it goes. In particular, think about the implications of your work in the real world. Could your work be used for weapons and war? It's surprising how many exciting breakthroughs in science and elsewhere could easily be turned into misery.

But it's not just the gigantic stuff. Your paper will have depth and verisimilitude if you've already answered the next several questions on the readers' mind.

WORKED EXAMPLE: *If the entire population is vaccinated, who is susceptible?*

This was a question I realised was dogging me with the infection spikes. Everyone was vaccinated, so how on earth did the disease spread? It took me a while to come up with the answer... but this was absolutely time well spent. And it's not optional. I guarantee someone else will ask this, and you'd better have a good answer.

Interruption: I'm just going to pause here and note — on advice of Anthony Wilson, my beta reader, who's not a mathematican in the slightest — that what's happening here is entirely deliberate. If you've been following along, you'll have been slowly gathering the tidbits of information about the infection spikes and why they matter, but not yet what's causing them, even though there are clues seeded in the narrative. Let me quote Anthony's comment in full:

> 'And *this* is making the narrative work! You tell us you have an answer — but not what it is. I think you should somehow highlight that's exactly what you've done here and relate it back to your earlier paragraph about plot twists, tension etc., as, if you're really writing for people who can't write, they won't necessarily spot that you've just done this by themselves. I should note that I am heavily invested in this background narrative that you have going on, which is how it ought to be working, which was why I was frustrated when I didn't understand the early parts of it.'

I'll note that this kind of feedback is utterly invaluable, because it tells me that what I'm doing has been noticed and has played out just as I wanted it to... while also giving me the idea to explicitly interrupt the book's narrative here to show you how a reader's comments can have an impact. Thanks, Anthony! Okay, back to the book.

WORKED EXAMPLE: *Is the vaccine realistic?*

Well, that one was easy: absolutely not. Indeed, I went to great lengths to stress that the parameters used for the infection spikes are extreme ones, not realistic. I wrote it in the Abstract. I wrote it in the Results section. I repeated it in the Discussion. I very much wanted to be clear that this was not a realistic case.

But I did more than that: I consulted with Geoff, my more biological co-author. I wasn't sure about using such weird values, even if they did give me unexpected results, so I asked for a second opinion. My co-author was equally intrigued, perhaps even more than I was, and he suggested that they worked as a proof-of-concept idea. That was invaluable.

What I was looking for here was the angle. It's not that the vaccine suddenly became realistic — it was never going to be — but rather that the weird parameters I was using performed another function: they showed that such a thing was at least possible, and that was of interest to biologists. The fact that Geoff was more interested in this than I had been at first helped enormously in giving me confidence that this was something people might actually want to read about.

WORKED EXAMPLE: *Why do we get these surprising results?*

Really, that's the question you should be asking every time anyway. Which presupposes that a) you have surprising results in the first place and b) that you understand why they occur. If you don't have surprising results, that just means you're not done researching yet, so keep working. And if you can't explain surprising results, they may be too good to be true. Better to find this out now than later.

Surprise is excellent. But then it's critical to understand the logic behind it. You'll be quizzed when you present this work (either by reviewers or by audience members when you present it as an academic talk), and you'll be invisibly quizzed about it by your readers, with no right of reply on your part. So you need to be able to justify it upfront, in the manuscript itself.

WORKED EXAMPLE: *Vaccinated individuals have a small chance of being infected. The transmission is very high, but the waning is extremely fast, so they don't stay infectious for long. The disease can thus act very intensely... but only in a small window. This is what causes the spikes.*

Ah-ha! Got it. This is a perfectly sensible explanation for the thing that had been bugging me. Everyone was vaccinated, but the vaccine isn't perfect (as no vaccine ever is), so that's where the 'susceptibles' are. And the super high transmission rate combined with the incredibly fast recovery produces the perfect conditions for spikes. I hadn't expected spikes at all, and certainly not in the case when everyone was vaccinated, but it now makes perfect sense that they'd be there.

The lesson here is: keep looking until you surprise yourself. Then work really hard to ensure that the surprise can be explained.

Interruption: Anthony commented on the above worked example, saying:

'And now the resolution of the plot! And I understand it!'

That's incredibly satisfying when that happens. Anthony gave me any number of notes about things to change or delete or move around, but he also tells me what's working, which I find equally as valuable. It means I'm not just writing in the dark. One of the downsides of writing is a disconnect from your audience: many people may read your academic paper, your novel, et cetera, but few will actually give you feedback on it — and those that do often nitpick it. Having a beta reader who can tell you what's working as well as what's not is utterly invaluable.

1.5.1 *Titles*

This surprise also is a great way of discovering the title.

WORKED EXAMPLE: *'Unexpected Infection Spikes in a Model of Respiratory Syncytial Virus Vaccination.'*

Yes, discovering. You want to find the title, not impose it on the article. The reason this title works is because it gives you the punchline (the infection spikes) but also tells you that it's not what you'd expect, while nevertheless communicating the essential information that this is a model and that it's about RSV.

The best titles are those that can intrigue and entice. They should get the reader wanting to know more. Crucial questions are How? and Why? But these aren't questions that go into the title; rather, these are questions that your title needs to inspire in the reader's mind. So make your title active, not passive.

BAD EXAMPLE: *'Comparison and validation of two mathematical models for the impact of mass drug administration on* Ascaris lumbricoides *and hookworm infection'* [Coffeng *et al.* (2017)]

I'm sure this was a perfectly fine paper, but the title doesn't inspire anyone to read further. What pressing question is being addressed here? Two models are compared; so what? That's an entirely passive title. Where's the intrigue and excitement?

WORKED EXAMPLE: *'Can we spend our way out of the AIDS epidemic?'* [Smith? *et al.* (2009)]

Well, can we? That's a big question. In fact, the key question here isn't 'Can we?' (since it's implied that we can); rather, it's the inferred 'How?' that immediately follows in the reader's mind. This example shows a much more active title than the previous one. Decide for yourself which one you'd rather read.

The same idea applies to the Abstract. We'll talk more about designing good Abstracts in subsequent chapters, but you also want your Abstract to entice your reader. (Remember Figure 1.2!)

BAD EXAMPLE: *'An ecological system consisting of tri-trophic food web system involving prey's fear is proposed and studied. It is assumed* *that the predation process throughout the system follows Lotka-Volterra type of functional response. The positivity, existence, uniqueness and boundedness of the solution are studied. All the possible equilibrium points are determined. The local stability analysis of each point is investigated. The persistence conditions of the system are established. The global stability of each locally asymptotically stable is carried out using Lyapunov method. The occurrence of local bifurcation around each equilibrium point is also studied. Finally, this paper is ended with doing numerical simulation to confirm our obtained analytical results and understand the effects of varying parameters on the global dynamics of the system.'*

This is an Abstract I was sent, and it's perfectly serviceable... but nothing here tells us why we should care. What's the central idea? It turns out there is one, but they buried the lede.

(Re)worked Example: *'A functioning ecosystem depends on the framework of its food webs. An underexamined aspect of such ecosystems is the prey's fear response to predation. We propose an ecological system consisting of a tri-trophic food web with a fear response for the prey and a Lotka–Volterra functional response for predation by both a specialist and generalist predator, which we call the superpredator. We prove positivity, existence, uniqueness and boundedness of solutions and determine all equilibrium points. We use Lyapunov functions to prove global stability, determine local bifurcations and illustrate our results with numerical simulations.* **Surprisingly, one outcome of the prey's fear of its specialist predator is the eradication of the superpredator.**' (emphasis added)

This was my reworking of the Abstract for the second draft. (It changed again before publication (Fakhry *et al.*, 2022), but only slightly.) The major change is the surprising outcome, which I've bolded. A prey being afraid of its major predator can eliminate an entirely different predator? That's huge!

I know I for one got super excited when I reached this point. Enthusiasm is infectious, even in the dry, dusty world of academia. We all know that the best teachers are the ones who radiate nerdish infectious enthusiasm for their subject. Conversely, if you can't be excited about your work, then why would you expect anyone else to be?

This breakthrough also gave the paper its title:

WORKED EXAMPLE: *'Prey fear of a specialist predator in a tri-trophic food web can eliminate the superpredator'*

Now there's a title and Abstract that's intriguing, that tells a story. Don't you just want to read more to find out how?

Chapter 2

The Introduction: The Goldilocks of Writing

2.1 Why have an Introduction in the first place?

At this stage, you should have all your basic research done and a draft of the Results. You should have a title and an idea of the shape of your manuscript. What you should not have at this point is the Introduction, the Discussion or the Abstract. These are generally things the reader will look at first. On no account should they be the thing you write first, or else it'll show.

More specifically, you need to know where the piece is going before you write the Introduction. If you think your article is going to be about treatment but further research reveals that it actually needs to be about vaccination, then you'll have just wasted valuable time, energy and words if you write all about treatment from the outset. Worse, once these words are written, it becomes hard to let go of them, and the end result is a Frankenstein's monster of patchwork pieces. You don't necessarily have to have finished all the work before you start writing, but you should at least have a rough guide of where it's going and what the topic is so that you can write the background.

As I'm writing this very book, I've thus far mapped out a plan, I have skeleton notes for each section and I've written everything up to this point... except the book's introduction. I'm saving that for later. Much, much later. I'll probably write it after I've done a first draft. Even though I know the shape of this book already, I'm still discovering small bits and pieces as I write that will reshape the direction. Best to wait until that creativity is complete before leading people into the book.

When it comes to the actual writing of your paper proper, the Introduction is the thing that most people will see first and hence judge you on.

It needs to guide them to a place where they're ready to receive your work. . . but you can't just start with that, any more than a joke can be only a punchline. You need the setup first.

A golden rule: Only things that are introductory should be in the Introduction. That sounds pretty obvious, but you'd be surprised by how often I've seen people throw their methods or results into the Introduction. That's not what it's for.

A corollary: Things that are introductory should not be in other sections. If you want to introduce something midway through. . . go back and put it in at the beginning. This is another good argument for not writing the Introduction until you have a roadmap, so you know what it is you're actually introducing. What we're crafting here is basically Chekhov's Gun, which is a rule for playwrights: a gun fired in Act 3 must be seen in Act 1; conversely, a gun seen in Act 1 must be fired by Act 3.

One thing that sets the Introduction apart from most of the other sections is that it needs to be tight. Every paragraph needs to contain a dense amount of information, sandwiched together beautifully in an elegant and efficient manner. Don't ramble. It also needs to be well-written. Of course, ideally, every section should be well-written, but this applies even more to the Introduction than other sections.

One of the perils of teaching is that the more expertise we have in a subject, the less we're able to come down to the level of someone who doesn't. They ran an experiment once, getting adults to draw houses like a five-year-old and then compared the average of these houses to the pictures that actual five-year-olds drew. Despite trying to come down to the level of a child, the adults drew vastly more complicated houses than the children did. . . with the exception of any adults who actually had five-year-olds. Likewise, your Introduction is supposed to appeal to a general reader but will likely overestimate their familiarity with your discipline. You need to consult with the equivalent of five-year-olds: namely, people outside your field of expertise, who are the audience for your Introduction.

2.1.1 *Acronyms*

Let's talk acronyms, a major bugbear of scientific writing. Here are a few do's and don'ts:

Don't overdefine basic terms (e.g., you probably don't need to explain what HIV stands for). The exception is if you're writing a thesis, in which case the Introduction is a lot more comprehensive.

Do define acronyms that aren't just basic terms. A good guide is: would a newspaper define the acronym? If not, then you don't need to. Otherwise, do so. (If in doubt, define.)

Don't define an acronym if you're not going to use it again; write it in full if it only appears once.

Do use acronyms consistently if you're writing for a specialised audience. Just be sure to define them early on if it's not patently obvious what they stand for.

Don't define an acronym more than once. (The exception is the Abstract, which you can treat as separate. If you need an acronym elsewhere within the Abstract, then define it both in the Abstract and also in the Introduction. If the acronym doesn't appear more than once in the Abstract, don't define it there but write it out in full until the Introduction.)

Do use the acronym after you've defined it.

Many Introductions in scientific writing end with an overview of how the manuscript is organised. These are pretty standard — but are they necessary? The answer is clearly yes if the organisation is confusing... but the other option is to consider resolving the confusion. Sometimes the standard structure is the best, and for most academic articles that's very easy:

- Abstract
- Introduction
- Methods
- Results
- Discussion

That's really all you need. (A few disciplines switch the order, such as putting Methods at the end.) Of course, several of these can have whatever subsections you want, but everything should fit into these overall categories. If your organisation is so intricate that you need to lay out a roadmap, maybe that's a problem.

As with everything, you can mess around with this structure, but you need to know the rules before you break the rules. My own feeling: don't put the organisation paragraph in. Instead, end the Introduction on a flourish, not a damp squib. Your readers will thank you.

BAD EXAMPLE: *'One of the most significant advantages of UAVs is their ability to work in complex scenarios and in regions where human* *operations are difficult. The real world scenarios can be visualize as 3-D arena for UAV. The 3-D path planning problem is categorized as the NP-hard problem with a primary purpose of creating optimal trajectory from the source to the destination. One of the essential tasks for UAV is to proceed optimally from a source to a goal while simultaneously avoiding other obstructions in the environment.*

There are different methods for path and trajectory planning which consist of the expectation of the complete path before beginning the movement (offline) or the case where the vehicle generates the path while moving towards the destination point (online). The 3-D path is represented as an arrangement of coordinated points.

Various meta-heuristic path planning algorithms in recent years of research are applied for finding a feasible path solution from the source to destination in a given environment. These algorithms produce more productive in comparison of proactive approaches and hence give most optimum paths in a complex domain with lesser execution times than those of the deterministic algorithms (1). These algorithms are very vigorous with wide range of applications and the capability to obtain high-quality solutions for numerous issues and situations.

Initially researchers use conventional method for path planning but it is less time efficient and then some researchers use meta-heuristic algorithms which give optimal solutions and then modifications in meta-heuristics are introduced to make them more useful for path planning application.

Although many meta-heuristics and deterministic algorithms exist, but their use for planning path of multiple UAVs in 3-D environments is not explored to the complete limit. In one recent work (2) MVO algorithm is used for path planning in 2D environment and good results are obtained. This inspired to select MVO for UAV path planning in 3-D and Munkres for coordination of UAVs in case of multiple targets.

Contribution of this paper included presentation of the approach for 3-D path planning based on MVO algorithm and comparison of the results of various cases of simulations with meta-heuristics Glowworm Swarm Optimization (GSO) and Biogeography-based optimization (BBO). For choosing algorithms to be compared with, the results the work (3) is viewed.

The originality of research work presented here in this paper is combined use of both probabilistic algorithms and deterministic algorithm for the path planning and coordination of UAVs in 3-D environment, respectively. MVO algorithm is used for generating path whereas Munkres algorithm is used for coordination purpose.

The paper is structured in form of different sections: Section 2 is about background and literature reviewed for understanding use of meta-heuristics in path planning. Section 3 illustrates the formulation of the problem and describes the environment for the problem of planning of path and associated parameters used. Section 4 describes MVO algorithm along with the customization done to it in order to generate optimal solutions. Section 5 explains the procedure and implementation to get the optimal path and to generate the trajectory. It also contains the experimentally obtained results by comparing with meta-heuristic algorithms such as BBO and GSO algorithm. Finally, this paper concludes in Section 6.' [Jain et al. (2019)]

There's a lot to unpack here. The most obvious is the abject failure to use the English language. The lack of articles or noun–verb agreements make this painful to get through and actively prevent the brain from assimilating the information. (The least helpful construction in the history of the written word may well be 'produce more productive', while the final sentence is inadvertently hilarious, essentially saying, 'Finally, we conclude.' Well, then!) Second is the failure to define acronyms (they aren't defined in the title or Abstract either). Third is the paucity of references, something we'll get to in Section 2.3.

Remember the skeleton idea from Section 1.3.2? Let's see if we can go backwards and construct the skeleton for each paragraph:

(1) Why UAVs are good (whatever they might be)
(2) Planning methods
(3) Recent planning algorithms
(4) A change in methods
(5) Planning paths haven't been fully explored/a recent publication had good results
(6) What this paper is doing
(7) Why these results are original
(8) How this paper is organised

Item #1 is good, if only we'd had an explanation of what a UAV is (with a reference). Items #2–4 should clearly be a single paragraph (with references to recent work). Item #5 is a mess; what is this paragraph actually about? Items #6–7 should be a single paragraph (although more properly moved to the Discussion). Item #8 is only necessary because the standard structure isn't being used, and I think that's detrimental to the paper.

Now, I'm not an expert in this field, and I chose this paper pretty much at random. But nothing about this Introduction inspires me to want to read on. I don't have a sense of what's come before that tells me that these results are going to be exciting. Instead, I'm confused by the organisation, something that could have been avoided with better planning or showing it to a non-expert first. A simple polish would do wonders here.

2.2 What to actually write

As you might have guessed from the previous example, the place to start is the skeleton. The good news is that this is really simple, because the Introduction is only a few paragraphs long.

WORKED EXAMPLE:

(1) *What is RSV?*
(2) *Immunity*
(3) *Seasonal oscillations*
(4) *Treatment*
(5) *Previous models*

See how easy that is? And if I decide to add or delete paragraphs, it's very painless at this stage. If I decide that two paragraphs need to be swapped, I have this view from 30,000 feet, instead of getting bogged down in the details.

Now all you need to do is to fill out each of these lines as a paragraph. As you do, try to find interesting ways to present the material. Your sentences should not all have the same structure. Some can be short. Others, if needed, can contain subclauses, which lengthens them, providing variation and depth. Or not. See what just I did there?

WORKED EXAMPLE: *'Respiratory syncytial virus (RSV) is the main cause of acute lower respiratory infections in infants and young children (22), with almost all children having been infected by two years of age (10,25) and an estimated 0.5–2% of infants requiring hospitalisation due to infection (18). One recent study estimated in that in 2005, 33.8 million new episodes of RSV occurred worldwide in children younger than five years of age (22). Symptoms of RSV range from those of a cold, more severe afflictions such as bronchiolitis and pneumonia (10). While mortality due to RSV infection in developed countries is low, occurring in less than 0.1%*

of cases (32), little data is published about RSV morbidity and mortality in developing countries (34). However, estimates of the hospitalisation costs are substantial (14,30,36), making RSV a significant economic and healthcare system burden.

Newborn infants are typically protected from RSV infection by maternal antibodies until about six weeks of age (8), and the highest number of observed RSV cases occur in children aged six weeks to six months (5,27). Immunity to RSV following an infection is short-lasting and reinfection in childhood is common (19). Few studies have been undertaken to investigate transmission of RSV among adults, but it is thought that infection can occur throughout life (6,15) and that in older children and adults, RSV manifests as a mild cold (10,16). RSV has been identified as a cause of mortality in the elderly with documented outbreaks in aged care settings (13,31); one such study found that up to 18% of pneumonia hospitalisation in adults aged above 65 years may be due to RSV infection (12).

In temperate climates RSV epidemics exhibit distinct and consistent seasonal patterns. Most RSV infections occur during the cooler winter months, whether wet or dry (34), and outbreaks typically last between two and five months (11,23). In a number of temperate regions a biennial pattern for RSV cases has been identified; see, for example, (4,20,28). In tropical climates RSV is detected throughout the year with less pronounced seasonal peaks, and the onset of RSV is typically associated with the wet season (26,34).

Immunoprophylaxis with the monoclonal antibody Palivizumab, while not preventing the onset of infection, has proven effective in reducing the severity of RSV-related symptoms (29). However, prophylaxis is expensive and generally only administered to high-risk children, with recommendations varying across jurisdictions. There is currently no licensed vaccine to prevent RSV infection, despite about 50 years of vaccine research. Recent research has focused on the developed of live attenuated vaccines; several such vaccines are being evaluated in clinical trials, with other vaccines in preclinical development (9,14). With the possibility of a RSV vaccine becoming available, mathematical models can be powerful tools for planning vaccination roll-out strategies.

Several ordinary differential equation Susceptible-Exposed-Infectious-Recovered (SEIR) type mathematical models for RSV transmission have been published to date, such as those presented in (3,7,17,21,24,33,35) with a sine or cosine forcing term to account for seasonal variation in transmission. Weber et al. (33) present a SEIRS model which incorporates a gradual reduction in susceptibility to reinfection and maternally derived immunity, and fit the model to several data sets. Leecaster et al. (17) present a SEIDR model with both child and adult classes for the S, E and I compartments, and where the D class represents children in which infection was detected. The model is fit to seven years of data from Salt Lake City, USA.

Moore et al. (21) present an age-structured SEIRS model for children under two years of age and the remaining population. The model is fit to data from Perth, Western Australia. Capistran et al. (7) outline a SIRS model with seasonal forcing and propose a method to estimate the model parameters, demonstrated by fitting models to data from the Gambia and Finland. Paynter et al. (24) investigate the ecological drivers of RSV seasonality in the Philippines, where the model includes a second partially susceptible class, and second classes for latent and infectious individuals with a subsequent RSV infection. This work also applies a square wave transmission term that accounts for decreased transmissibility over the summer holidays, as well as a seasonally driven birth rate.

White et al. (35) describe nested differential equation models for RSV transmission and fit these to RSV case data for eight different regions. In the work of Arenas et al. (3), randomness is introduced into the differential equation model and the model fit to RSV hospitalisation data for Valencia, Spain.

Few papers have so far explored vaccination strategies for RSV. A new-born vaccination strategy is outlined in (1) for the Spanish region of Valencia, in order to estimate the cost-effectiveness of potential RSV vacci-nation strategies. The modelling approach removes a fraction of susceptible newborns into a vaccinated class, where they remained until they reached the next age group, at which point they move to the second susceptible class. This strategy assumes booster doses of the vaccine in the first year of life, such that the immunisation period would be at least equal to the immunity of those who have recovered from RSV infection. In subsequent work, an RSV vaccine cost analysis is conducted based on a stochastic network model, with children vaccinated at two months, four months and one year of age (2).

Details of what we plan to do...'

Notice how this follows the skeleton structure, how every sentence is referenced and how the first four paragraphs are, with one exception, tightly written. You'll also notice that the paragraphs on previous modelling aren't tight; they need to be condensed into a single paragraph (and will be), but for now it's more important to get the information down. The final line is a note from my co-author suggesting an additional paragraph: one that outlines what we're planning to do, which was a good suggestion. The exception is the 'symptoms' sentence. See Chapter 5.

The Introduction is the Goldilocks of writing: not too little, and not too much; just right. This can take a bit of experience to figure out, so find the well-written papers in your field — by now, you should be able to judge which ones they are — to get a sense of what they're like. But you definitely want to include all the pertinent information, without rambling. Get that balance right, and your paper is off to an excellent start.

The aim of the Introduction is to situate your work within the context of what's come before. You need to tell the reader what work has been done in the field, what the problem is, what the techniques are that you'll be using and so forth. Be sure to acknowledge the shoulders of giants you're standing on. But also keep it tight.

2.3   References: How many and which ones?

You'll note from the previous example that there are over 30 references here. That's a pretty good guide. Usually, you want 20–40 references in total, the majority of which should be in the Introduction. (Some might be for data that are found in a table.) Generally, if you're putting a new reference in later in the paper, ask yourself why that is and whether that part should really be in the Introduction. At the very least, there should be a good reason for it not being there.

Also, (almost) every sentence in the Introduction should have a reference. This is where the UAV Introduction falls down, because a lot of claims are made without any evidence to back them up. The Introduction is not your work; that's what the rest of the paper is for. So justify any claim you make.

Which references should you use? Only peer-reviewed articles. Except for raw data, don't use websites. Don't use pre-reviewed repositories like arXiv. Anyone can post anything up there, and without peer review, it could just be nonsense. Academia can be slow and stately sometimes, but that's because it takes time to verify claims and craft the work into shape.

A note for mathematical biologists: references concerning biology should be to the biological literature, not to other mathematical models. This probably generalises to other interdisciplinary fields, but make sure you actually go back to the source, even if it's less familiar ground. I've seen too many modelling papers cite data from other modelling papers, which turn out to have no legitimate original source.

Be consistent in your citation style. Google Scholar is a great resource and has a ⌜"Cite⌟ button that gives you a variety of styles. Pick one and stick to it. It doesn't have to be the style of the journal you submit to (they'll adjust it later or get you to), but it's best to be consistent when sending things out for review.

If you're writing in LaTeX (and who wouldn't want to, it's wonderful!), then a BibTex file is an excellent resource (rather than in-document citations that have to be adjusted manually). Start collecting now; your future self will thank you. One line of code at the beginning of the document can adjust to any citation style if you collect the information this way. Google Scholar has copy-and-paste options for BibTeX, EndNote and a few others, which makes life very simple. However...

...**always** check with the original source. The metadata that goes into Google Scholar isn't always accurate and is often very inaccurate. I say this as someone whose surname is consistently misspelled there. But it's not just my weird punctuation; volumes and page numbers and even journal titles can be off. So use this excellent resource for a first draft, but make sure you double check it with the original before sending it out for review. You'll look like an idiot if you don't.

BAD EXAMPLE : *'Diphtheria and tetanus are both vaccine-preventable diseases, and, in China, are reportable infectious diseases. The main vaccines used in the prevention of diphtheria, tetanus and pertussis are DTwP or DTaP; the main components of these vaccines are diphtheria toxoid and tetanus toxoid combined with either whole-cell pertussis (DTwP) or acellular pertussis (DTaP). The basic vaccine DTP has also been combined with Haemophilus influenzae type b or inactivated polio virus to create multivalent conjugate vaccines. It is known that DTaP vaccines have a lower efficacy than DTwP vaccines against pertussis (1,2), but little is known about the long-term efficacy of DTaP against diphtheria and tetanus.*

In China, both DTwP and DTaP vaccines were licensed by China Food and Drug Administration (CFDA). DTaP vaccine in China was produced with the method of immunoprecipitation (3). Primary vaccination of infants and young children consists of three doses of DTP (DTwP or DTaP) in the 3rd, 4th and 5th months of life followed by a fourth dose at 18–24 months, with Diphtheria-Tetanus vaccine (DT) booster doses being recommended at 7 and 12 years. DTwP was introduced in 1978 and was replaced by DTaP in early 2007 because of its potentially severe adverse reaction. In China, primary vaccination with DTP (DTwP or DTaP) is mandatory, while the DT booster doses are not. According to official estimates, since 2002 the immunization coverage rate achieved with three doses of the DTP vaccination in childhood has been more than 90% (4). In 2011 the immunization coverage of four doses was over 99% (5). In Gaobeidian city and the first generation of children receiving solely DTaP as their primary vaccination is now aged 6 years.

Antibodies against diphtheria and tetanus toxins decline over time. In children aged 5–9 years antibodies against diphtheria and tetanus are low if the booster dose is not given until they are 10 years old (6–9). However, an evaluation of the antitoxin against diphtheria and tetanus has not been carried out in Chinese children and adolescents. After the introduction of DTaP, a world-wide efficacy evaluation of DTaP was conducted, but this focused only on the sero-conversion rate after vaccination (10,11). In China, little is known about the long-term changes in antitoxins against diphtheria and tetanus after the replacement of the DTwP vaccine with DTaP.

This is a cross-sectional study that measured the levels of IgG antibodies against diphtheria and tetanus. The aim was to determine the sero-epidemiology of diphtheria and tetanus among children aged 2–6 years old and students aged 7–17 years old in Gaobeidian city, China.' [Wu et al. (2014)]

The organisation here is pretty good. Try writing out the skeleton of this Introduction if you want some practice. However, the same problem of not defining acronyms comes with up DTwP and DTaP. They're clearly important but are never defined. And there's a serious lack of references: only eleven in the entire Introduction. Every sentence in that first paragraph should — and easily could — have a different reference.

It follows that all the working parts of the Introduction need to be functioning together. The skeleton, the prose and the references all matter, and they all need to be checked and crafted.

BAD EXAMPLE : *'Leptospirosis can be defined as a zoonotic bacterial caused disease, considered as one of the most geographically widespread zoonosis in the world with greater incidence in the tropical region (1). Leptospirosis is caused by spirochaetes of the genius Leptospira and affects millions of people worldwide charan et al (2), it is a major cause of disease in domestic animals. Leptospira are conventionally divided into species, the pathogenic Leptospira interrogans and saprophytic Leptospira biflexa faine and stallman 1982 (3). There are over 300 serovars and 25 serogroups G. de Vries et al (4). Different hosts carry distinct serovars (1). Leptospires were among the least studied micro-organisms in medical microbiology largely due to practical difficulties in the laboratory examination. However, Leptospirosis has gained international recognition of late, this is as a result of its outbreaks in Asia, the United states, central and south America, owing to flooding related to excessive rainfall and climate change. Moreover, in Africa, countries in west Africa, central Africa, East Africa and South Africa have recorded incidences of leptospirosis but little effort has been done see (4).*

The transmission of infectious disease from wildlife to humans represents a serious and growing public-health risk owing to increasing contact between humans and animals. This study exploit the use of impulsive differential equations to study new avenue for the spread and the control of leptospirosis. Our work is the first since the mathematical model for the study of lepstospirosis was develop in 2006 (5), to use impulsive differential equations to study the transmission and treatment of leptospirosis. (6)

Modelling with impulse has been used as an important tools in the controlling of diseases such as HIV/AIDS (11), Malaria (12), Guinea Worm (13).'

Let's start with the good: The first paragraph is a great overview of what leptosirosis is. The references here are pretty good, although they do cluster weirdly at the end and skip over a few, and there's a strange numbering glitch.

But now the bad: The writing is atrocious (though this is just a first draft, and drafts don't get published, remember?). The Introduction is missing background material on intervention methods and mathematical techniques. A skeleton would have made these omissions really obvious.

To conclude our introduction to the Introduction, eventually, after all the fine-tuning that will come later, you will have something that you're prepared to put out there. Before you do that, get some other people to read it. If English isn't your first language, get a native English speaker to read it. In fact, do this even if English is your first language, because you're probably too close to it. Second, get a non-expert to read it. This can be your mother, your roommate, your best friend. More eyes on the words will give you more feedback about what works and what doesn't.

The great thing about the Introduction is that it's designed for people who aren't experts, so if they can't follow what you're talking about, then you need to rewrite it.

At this stage, we're more focused on the organisation and getting all the ducks in a row than we are on fine-tuning the words. That will definitely come later, but for now it's okay to have something rough. There's a lot of revision still to come.

Chapter 3

Collation: Filling in the Gaps

3.1 Finishing the research

It's now time to fill in any of the gaps in the body of the manuscript itself. Things you've pencilled in to do later... well, now's the 'later'. If there are any pieces of the jigsaw missing, then put them in place. It doesn't have to be pretty, but it does have to be finished.

WORKED EXAMPLE: *'For the RSV manuscript, this means doing the section on the impulsive analysis. For this, I made some overestimates and was able to find bounds on the impulsive orbit in terms of susceptible and vaccinated individuals.*

What I wasn't able to do was to find bounds for the infected individuals. I had to make a further assumption here: namely, that the number of people who were both vaccinated and infected was negligible. I wasn't very happy with this, as I don't think that's likely to be 100% true, but it was the best I could do. It's important to be ethical about this sort of thing, so be sure to mention such limitations explicitly.

Along the way, I came up with something pretty funky: an impulsive analogue of the reproduction number, which I called T_0. This incorporated regular pulse vaccination into the derivative of the infected compartment. I needed this to show that the disease would be controlled when $T_0 < 1$ (although the reverse wasn't necessarily true, as I had no results for when $T_0 > 1$).

I also added a diagram to illustrate the impulsive part. See Figure 3.1. This is a perfectly fine figure, showing the discontinuous jumps in the susceptible and vaccinated classes, along with the resulting effect on the other classes. However, I wasn't super excited by it, as it's very standard in the field. Sometimes that's enough.'

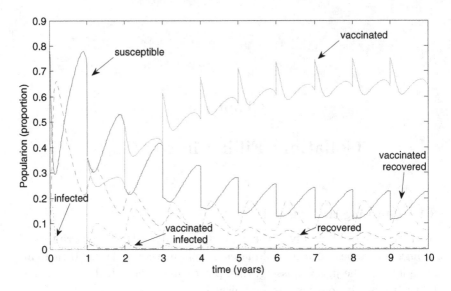

Fig. 3.1. Impulsive jumps and their effect on other classes.

In some ways, the impulsive stuff wasn't all that exciting. That had been my original idea, but it just didn't sing the way the infection spikes had. No matter: I already had some good stuff, and this needed to be added. However, the impulsive reproduction number was something pretty amazing. It's an analogue of the basic reproduction number that's used all the time in disease management, but with impulses incorporated into it. I've spent my career investigating diseases and impulses, but this was a way in which the two areas really came together. Even better: I thought I might use this again (or others might), so it was definitely worth highlighting. The fact that something so exciting came out of a fairly bog-standard investigation was really cool.

The key here is to do the grind of anything that has to be done, whether you feel like it or not. Remember: everything will be revisited later on. So you don't need to have it all perfect at this stage. But what you do need to do is have it down on the page. If you don't write it down, you didn't do it.

3.2   Should you add a co-author?

At this point, it's worth addressing writing fatigue, which can sometimes set in. It's all too easy to feel demoralised or as though your results aren't as exciting and original as you might like them to be. Or that the words sound tone deaf when they should be singing. We all get imposter syndrome, and the biggest danger can be simply abandoning work that might otherwise be good, except that we get too close to it.

A solution to this problem that I've found is to add a co-author. It's a great trick that can save a flagging paper. Someone from outside might be quite happy to come on board to a paper that's 75% done, even if that final 25% looks impossible from your exhausted perspective. I've done this several times, and it injects new energy into the project. A new person often brings new insights and enthusiasm, which can help massively.

WORKED EXAMPLE: *'For the RSV project, I had a much sadder reason for bringing in a co-author. My collaborator Geoff passed away unexpectedly during the process. He was a well-respected researcher and an excellent collaborator and colleague, whom I still miss to this day.*

Geoff's passing left a lot of people in the lurch, including his students. One of them, Alexandra Hogan, had a lot of biological expertise in RSV, so I invited her to join the project. I missed Geoff's insights and still needed the perspective of someone better versed in the biology than I was. Alexandra was exactly that person.'

Regardless of the reasons, there's basically no penalty for adding a co-author, at least in the sciences. Many biological papers have multiple co-authors on them, so adding another doesn't cost anything. The exception is in a few sub-fields where solo papers are viewed more favourably. However, even here I'd argue that sharing some of the glory is definitely preferable to getting no glory. Better to have a co-author bring your project back to life than be the solo author on a never-submitted manuscript.

Finally, if English isn't your first language (or even if it is, but writing isn't your thing), consider adding a co-author who can help with this. I've been added to many, many projects over the years for this reason. Of course, I bring my mathematical expertise as well and examine the logic behind all the derivations... but it's in the rewriting and editing that I usually contribute most, because many people don't have these skills at all. If this is a weakness of yours, consider ways to turn it into a strength. Adding someone who can liven up your writing can be a great way of getting more attention for your work.

3.3 The Discussion

It's only now, when all the research is finished *and* written down, that you start the Discussion. There's absolutely no point doing it beforehand, because you don't know what's actually in your manuscript — and what lies beyond it — until you've actually finished the thing. You can't have a Discussion until there's something worth discussing.

I'd like to address a common misconception here. The Discussion is absolutely *not* an extended summary of your results. By all means summarise what you have — in two or three paragraphs, no more. And that should be less than half of the Discussion. You don't want to spend the bulk of the Discussion talking about what we've just read. That's redundant and also not what the Discussion is actually for.

Instead, the Discussion is for implications beyond your work. Where might these results lead if one were to continue? What are the potential applications? You should also state any limitations in your original premise, both implicit and explicit.

It's up to you if you want to discuss future work. I sometimes do and sometimes don't. The future-work section can be a good way of pinning down ideas that you might come back to later. On the other hand, I've often written down future ideas and never looked at them again, so I don't always bother with this. (Which isn't to say that other researchers won't pick them up, of course!)

Finally, your concluding paragraph should be solid and contain some recommendations. Make sure any such recommendations are biologically sensible. Make sure you finish with a bit of a flourish. Far too many papers just sort of... stop.

WORKED EXAMPLE: *'The introduction of a vaccine is always desirable, but new vaccines pose the risk of unintended consequences. We have highlighted some of the potential issues that may arise with vaccination against RSV. In particular, we determine conditions under which a destabilisation of the disease-free equilibrium is possible. This is not in the form of a backward bifurcation, as is sometimes seen, but rather occurs when the vaccine causes sufficiently fast recovery and transmission is extremely high. An infection-free population that is effectively vaccinated against RSV can nevertheless produce vaccination-innduced [sic] spikes of infection. Although such a case is unlikely to occur with the unrealistic parameters we chose, we have shown proof-of-concept that it is possible and determined* *conditions on the recovery rate due to vaccinaton [sic] that allow for the possibility.*

Using impulsive differential equations, we were able to formulate conditions on the period and the strength of vaccination to allow for disease control (though not eradication). If the vaccine reduces transmissibility and is applied frequently, then vaccinated infected individuals can be reduced to low numbers. We relaxed the assumption of constant transmission. We demonstrated that the waning of the vaccine has a greater effect on the outcome that [sic] coverage. Hence it is imperative that a good vaccine be developed before being released for general use.

We also defined a new quantity, the impulsive reproduction number T_0. This is a sufficient (but not necessary) condition, based on an overestimate of the infected population, that ensures eradication if $T_0 < 1$. If $T_0 < 1$, then the infected population is contracting within each impulsive cycle. Since the infected population is then reduced at each impulse point, the result is the eventual eradication of the infection. Note that we assumed constant transmission for this derivation; however, numerical simulations were performed using seasonal oscillations. The result was a double period: one from the impulsive periodic orbit and the other from the seasonal oscillations.

Our model has some limitations, which should be acknowledged. We assumed that time to administer the vaccine was significantly shorter than the time between vaccine administrations in order to justify the impulsive approximation. Such assumptions are reasonable in many cases (30), although can produce confounding effects in some situations (11). The extreme parameters that we used to illustrate the vaccination spikes operated under the assumption that the transmission rate for infected vaccinatied [sic] individuals was significantly higher than the transmission rate without vaccination. Since we extended the model of Weber et al. (39), our model inherited many of the assumptions from that model, such as mass-action transmission, a constant birth rate and that the birth and death rates were matched, resulting in a constant population.

A vaccine that targets RSV infection has the potential to significantly reduce the overall prevalence of the disease, but it has to be sufficiently long-lasting. Coverage and effectiveness of the vaccine is important, but the critical parameter that our modelling identified is the waning rate of the vaccine. We thus recommend that candidate vaccines be tested for sufficient duration before being released to the public. If a durable vaccine can be developed, then we stand a chance of controlling this disease, assuming sufficiently widespread coverage.'

This was the original draft of the Discussion (typos and all) that I wrote for the RSV paper. Let's break this down by paragraph.

Paragraph 1 is an overview. The first sentence is a general call to action. The rest is a summary of the spikes, along with a note that they are based on unrealistic parameters, because I very much wanted to highlight the theoretical nature of this part. (I've left the 'innduced', 'vaccinaton' and 'vaccinatied' typos here, as obviously these will get corrected later.)

Paragraph 2 lists key results. This summarises the bread-and-butter of the manuscript, the impulses and the waning.

Paragraph 3 talks about a generalisation that extends beyond RSV. It highlights the new concept introduced along the way, which I want to flag because I'll likely be using it in future papers.

Paragraph 4 states the limitations, but also interrogates and in some cases justifies them. Some of these were explicit when I introduced the model, such as the extreme parameters, but others are implicit, such as the impulsive assumptions or the properties of mass-action transmission. Note that there are minimal references in the Discussion and no new ones. (If new ones are needed, that section should probably be moved to the Introduction.)

Paragraph 5 has a very brief summary and lists recommendations.

However, I want to stress that this is not at all complete. It's a first draft, and we'll be revisiting it as we go. For one thing, there's too much summary and not enough implications. There's not enough linking to the biology. For a first draft, it's okay. But I'd be hugely embarrassed if it were published as-is.

BAD EXAMPLE : *'In this paper, we investigated the transmission dynamics of the novel corona virus pandemic via an epidemic model. We formulated the model and studied the basic fundamental biological as well as mathematical properties for the well possed-ness of the problem. The threshold quantity of the model is calculated to perform the local and global dynamics of the proposed around disease free and disease endemic equilibria. We estimated the values of the epidemic parameters from real data of the novel corona virus as reported by the government of Khyber Pukhtunkhwa Pakistan from 13 April to 9th June 2020. On the basis of estimated parameters value, we discussed the detail local as well as global sensitivity analysis to find the role of every parameter on the disease transmission. Our* *analysis shows that the disease transmission co-efficient (β_2), the moving rate from asymptomatic to infected (γ_1) and the ratio contribute the virus to the seafood market from asymptomatic individuals (η_1), are very sensitive and statistically significant parameters. It may be noted from the sensitivity analysis that increasing the transition of individuals from the asymptomatic to the confirmed infected individuals causes reduction in the number of new cases. Finally the long run of the model are executed which shows the future forecast of the disease transmission. This describes that the novel corona disease will take time of more than a year for control and elimination.'*

This was the entire Discussion from the Pakistan COVID model, also with typos intact. As a summary of the work done, it's accurate enough. Why the variables are explicit, I have no idea. (Any non-mathematicians reading this who may have skipped the details are going to be put off by the Greek letters.) And the final sentence — the prediction that COVID-19 would run for more than a year — was quite amusing, given that I received this in October 2020, and the pandemic showed no signs of slowing, 10 months in (and if it were accepted, then the actual publication date would be many months hence). So why did we need a mathematical model to tell us it would run for at least a year?

Interruption: Anthony had a helpful observation at this point:

> 'On a related note, is it worth re-emphasising your position on the skeleton and paragraph breakdown again here, because I think this is one of the (numerous) weaknesses of your chosen bad example; I can't make my way through the text because of its lack of organisation?'

There's only one paragraph here, so what's its skeletal structure? My best guess is 'Here's what we did.' That's not very useful, though, nor is it what the Discussion is for.

3.4 The Abstract

Throughout this book, I've often stressed that my recommendations are guidelines, and that if you want to do things another way, then by all means do, so long as you know the rules before breaking them. I think this should probably happen a lot less than it does, and it should only be done if you *really* know what you're doing, but you are free to go your own way.[1]

---

[1] Though that does mean you face the consequences of those choices, of course.

However, I'm going to make an exception for the Abstract. *Do not mess with the structure of the abstract.*

What's the structure? Easy: Background, Methods, Results, Conclusion. This can easily be summarised as follows:

- *Background*: What is the problem? 1–2 sentences explaining why we should care, which sets the scene.
- *Methods*: How did you approach it? 1–2 sentences on your methods.
- *Results*: What did you find that was new? 1–2 sentences explaining the original findings that come out of your work.
- *Conclusion*: What are the implications? 1–2 sentences, though usually one is enough here.

Note the very limited number of sentences. And I don't mean run-on sentences, either. You want to keep each section short, sharp and focused.

Some journals will explicitly make you write in these categories. But even if the journal doesn't require it, you should do this anyway. I always do and comment the category out (using the % signs so they don't show up in the pdf), but you should absolutely have each of them there, regardless. Unless the categories are explicitly laid out, the Abstract will be a single paragraph. The reader should still be able to tell which parts belong to which category.

Worked Example:

%*Background:*

'*Respiratory Syncytial Virus (RSV) is an acute respiratory infection that infects millions of children and infants worldwide. Recent research has shown promise for the development of live attenuated vaccines, several of which are in clinical trials or preclinical development.*

%*Methods:*

We extend an existing mathematical model with seasonal transmission to include vaccination. We model vaccination both as a continuous process and as a discrete one, using impulsive differential equations.

%*Results:*

We develop conditions for the stability of the disease-free equilibrium and show that this equilibrium can be destabilised under certain (extreme) conditions. Using impulsive differential equations and introducing a new quantity, the impulsive reproduction number, *we determine conditions for the period and strength of vaccination that will control (but not eradicate) RSV.*

%Conclusion:
The waning rate of the vaccine is a critical parameter for long-term reduction in RSV prevalence, even more than coverage. We recommend that candidate vaccines be tested for sufficient duration before being released on the market.'

This was the first draft of my Abstract, written after all the research was complete. It's not quite the final version (as we'll see going forward — or else you can just google the published paper and compare (Smith? *et al.*, 2017)), but it's pretty close. I've highlighted the four categories, whose headings are usually commented out (with % signs).

The two Background sentences explain what RSV is and set up the idea of vaccination as something promising. The Methods sentences tell us about the extension from the original model and the addition of impulses. The Results sentences give us both the spikes and the impulsive reproduction number but also more biologically useful things like the stability criteria and the vaccine conditions for control. Finally, the Conclusion sentences highlight the waning as a critical parameter and make a recommendation for vaccines that may currently be in testing.

Abstracts are an excellent lesson in compact writing. As academics, we often go on tangents, which is the nature of research. What we're not always so good at is looking at the big picture. The Abstract forces you to do that, by virtue of both the formal structure and also the limited wordcount. These are good constraints to have. Being a writer means being able to adapt to the audience and to write to requirements.

BAD EXAMPLE : *'In developing countries, HIV is often managed not in hospitals but in community-based organizations. This includes access to medicines for pregnant women and people living with HIV through th CBOs.*

Once diagnosed, infected people can be sent to the community, on treatment or hospitalization.

Individuals can also go through these steps. People taking drugs can of course develop drug resistance, whether in the community or in hospitals. People may also stop taking medication or be discharged from hospital.

This thesis explores the effects that CBOs have on disease management in the community, which also has economic value, as significant hospital costs can be saved.

We propose models based on current knowledge about HIV transmission and treatment, allowing public-health awareness campaigns.

The aim is to study the mathematical modeling of the evolution of HIV/AIDS infection in the community, with a view to providing early support to the clinical diagnosis. Our approach is to model the different stages of viral treatment, reducing the rate of spread of the disease in a community for a healthy next-generation.'

I think the first thing that jumps out here is that there's way too much Background. Six sentences of background is just far too top-heavy. The other thing that jumps out is that this Abstract was clearly written before the project was finished. There's essentially no Methods, more a wistful hope, with only a vague sense of how this will be done. The structure is messed up, and it really hurts the piece. And the Results and Conclusion are entirely absent. (To say nothing of the appalling grammar or the weird paragraph breaks, which are entirely unnecessary, although in a published piece they would all be combined into one.)

Remember that most people will only read the title and the Abstract, so you need to tell the entire story in these eight sentences, across four categories. But, as always, don't ramble. The story needs to be tight, and it needs to intrigue the reader so that they are inspired to learn more and hence read the rest of the manuscript. Choose your words carefully and sparingly.

Okay, so with the Abstract written, your paper is complete! Right? Well, a first draft is. But this isn't the end of your paper, it's only the beginning. Now the real work begins...

Chapter 4

Refinement: This Is Where Your Manuscript Truly Begins...

> "The aim is not simply to make it possible for the reader to understand; the narrative should be strong enough to make it impossible for the reader to NOT understand."
>
> — Glasman-Deal [2020]

4.1 Editing

Remember the novelist who would rewrite each page taped around his house? This is where you start doing that. Of course, in a smaller work than a novel, the unit is not the page but the paragraph. So you need to be able to (1) assess the quality of each paragraph and (2) raise its standing. Neither of these are trivial skills.

As I mentioned earlier, all writing is editing. That's an absolute truism, and the reason we don't publish first drafts. I'm going to be more specific here: all writing is problem-solving. What are the problems in your manuscript? How will you solve them? This is where the hard work starts. It's time to start raising those paragraphs.

To identify problems, you need cast a critical eye over your work — and someone else's critical eye as well. To do this yourself, you need to possess said critical eye in the first place. If you don't, then find someone who does and get them to do it. Bribe them with snacks or co-authorship if you need to. Or offer to do the same for their manuscript (or secret novel that they're writing if it's a relative). But don't skimp on this step.

Valid choices for beta readers include a co-author (who should be doing this anyway), friends, colleagues, relatives. In fact, you probably want a mix of experts and non-experts to look at it, especially when you're starting out. Your manuscript should work for both groups.

I recognise that it's not easy simultaneously writing for both the expert and the lay person. Tough. This is your job. I'm very happy to give you the tools to figure this out, but you don't get to ignore this aspect. If a lay-person reads your manuscript and is completely bamboozled, you will have failed. If an expert reads your manuscript and is confused or bored, you will have failed. Both things are true.

If you have to do this yourself (and it's an excellent skill to possess), then one of your best weapons is time. Writing something and then immediately editing it is useless, because you won't have the required distance from it. By contrast, if you read back over something you wrote from five years ago, you'll see all the flaws with a — hopefully — dispassionate eye.

Don't cringe if what you find is horrible. Rather, celebrate the fact that this means you've grown as a writer! But you should be able to see what works and what doesn't with a clear head. The aim now is to do this in a shorter timespan.

I recommend at least two weeks. By which I mean two weeks after you last put down the manuscript and *do not touch it again.* You'll need to build this time into your schedule, so plan ahead. And if for some reason you go in and tweak something... well, the clock resets at that moment.

During those two weeks, work on a different project. And have some fun. Do whatever you need to distract your brain from the manuscript at hand, because you need to fool yourself into forgetting it.

Once you're ready to look at it, be as critical as you can. Find all the problems. What are the gaps in logic? What's missing? What is clear to you but not the reader? Tear this thing apart.

Specifically, what you're trying to do is find all the problems a reviewer would notice. And don't fool yourself into thinking there won't be any, because there will. Reviewers are excellent at spotting very fundamental flaws in your construction or holes in the logic or bits that simply don't make sense. The idea is to locate all these yourself (or with help of beta readers) so that the problems that the reviewers eventually spot will be small things in a corner that can be dealt with easily.

In a pinch, you can use the reviewers as invisible co-authors, but if you do this, then expect your process to be a *lot* slower. You'll be submitting much earlier versions than you otherwise should, so take all the comments into account every time you iterate the process. I had to do this when I was a postdoc in a non-math department and didn't have any colleagues who could read my work, but it meant each manuscript got rejected again and again until eventually I could get it right.[1] That's an experience I definitely don't recommend.

---

[1]My record is 21 rejections before acceptance. Nobody said being an academic was all adoring crowds and international fame.

Clarity is utterly crucial. This is especially true if your academic field is one that isn't strong on clarity to begin with, such as mathematics. Math is precise and careful, but it doesn't lend itself to easy reading, so that means you need to work doubly hard to communicate what you want to say to an audience that might be quite hostile to reading it in the first place. The same goes for other disciplines, though. If your jargon is getting in the way, strip it down to the essentials.

The next example is Alexandra's edits on my previous draft. As mentioned earlier, she was Geoff's student, but I'd never met her before Geoff died, so we had no pre-existing working relationship. This gave her some very helpful distance from the project. Her changes and suggestions are highlighted.

WORKED EXAMPLE: *'Newborn infants are typically protected from RSV infection by maternal antibodies until about six weeks of age (9), and the highest number of observed RSV cases occur in children aged six weeks to six months (6, 33). Immunity to RSV following an infection is short-lasting, and reinfection in childhood is common (22). Few studies have been undertaken to investigate transmission of RSV among adults, but it is thought that infection can occur throughout life (7, 15) and that, in older children and adults, RSV manifests as a mild cold (10, 18). RSV has been identified as a cause of mortality in the elderly, with documented outbreaks in aged-care settings (13, 37); one such study found that up to 18% of pneumonia hospitalisation in adults aged above 65 years may be due to RSV infection (12) *could take out this paragraph? .*

Immunoprophylaxis with the monoclonal antibody Palivizumab, while not preventing the onset of infection, has proven effective in reducing the severity of RSV-related symptoms (35). However, prophylaxis is expensive and generally only administered to high-risk children, with recommendations varying across jurisdictions. There is currently no licensed vaccine to prevent RSV infection, despite about 50 years of vaccine research. Recent research has focused on the development of particle-based and subunit vaccines; several such vaccines are being evaluated in clinical trials, with other vaccines in preclinical development (28, 30). With the possibility of an RSV vaccine becoming available, mathematical models are powerful tools for assessing the impacts of different vaccine characteristics.

Several ordinary differential equation mathematical models for RSV transmission have been published to date, most using Susceptible–Exposed–Infectious–Recovered (SEIR) dynamics and with a sine or cosine forcing term to account for seasonal variation in transmission (3, 8, 19, 24,

*27, 39, 41). Few papers have so far used dynamic models to explore vacci-
nation strategies for RSV, and these have generally investigated RSV
vaccination from a cost-effectiveness perspective (5, 21), for example in the
context of a newborn vaccination strategy in the Spanish region of Valencia
(1, 2). More recent studies conducted for the setting of rural Kenya have
focussed on the likely benefits of vaccination for particular target groups
(17, 29). *A nice segue here would be to say that we have not identified any
RSV vaccination models that examine the impact of a theoretical vaccine
analytically, and look at the stability of different scenarios - but can't think
right now how to word this.*
   *Here, we examine the effects of a theoretical vaccine on the transmission
of RSV in a single age class. We consider several vaccination scenarios,
including differing levels of coverage, seasonal oscillations in the transmission rate and a waning of the vaccine. We also compare continuous
vaccination to impulsive vaccination in order to determine conditions on
the vaccination strength and duration? that will control the virus.'*

   As is clear, Alexandra suggests big cuts, like taking out the entire
newborns paragraph. That was a surprise to me, because I thought it was
crucial, but the fact that she flagged it suggested there was a problem. It
doesn't mean that this paragraph was necessarily the issue, but it did mean
that I had to do something about it.

   I'll also note at this juncture that the prospect of cutting an entire
carefully referenced paragraph that I'd spent time on was painful. Had
this been a cut made at the skeleton stage, it would have been simplicity
itself to remove. Making global changes at this stage is still useful, but
it's harder to do. However, don't be afraid to make them if that's what's
needed.

   Alexandra added useful details to the paragraphs on immunoprophylaxis
and pre-existing models, which were a huge boost to the Introduction. They
add clarity and perspective, along with new research that's come to light.
Note her segue at the end, where she made a suggestion for better flow but
wasn't sure how to make it work. I took one look at that and knew exactly
what was needed, but I wouldn't have been able to do that without her
flagging it.

   In the final paragraph, she makes a few minor additions for clarity. All
of these were utterly invaluable. Some simply added, some questioned what
was there, and some pondered possibilities. This is exactly the benefit of

bringing in a beta reader with a fresh eye. (My responses are in the next section.) Here are Alexandra's edits to the Methods section.

WORKED EXAMPLE: *'We extend the basic model from Weber et al. (39) to include a vaccine strategy for RSV where a fixed proportion of newborns are vaccinated?. We assume that the leaving rate μ is unchanged across all classes and that there is no disease-specific death rate. We scale the entry and leaving rates so that the population is constant.*

Let S represent susceptible, I represent infected and R represent recovered individuals, with V, I_V and R_V the corresponding compartments for vaccinated individuals. The birth rate is μ, with a proportion p vaccinated, of whom ϵ successfully mount an immune response; the death rate is matched to the birth rate. The time-dependent transmissibility ~~parameter~~ function is $\beta(t)$, with recovery ν and loss of immunity γ. Corresponding vaccination parameters are $\beta_V(t)$, ν_V and γ_V, respectively. Finally, the waning of the vaccine protectiveness? is given by ω.

The basic model with vaccination is then

$$S' = \mu(1 - \epsilon p) - \mu S - \beta(t)S(I + I_V) + \gamma R + \omega V$$

$$I' = \beta(t)S(I + I_V) - \nu I - \mu I + \omega I_V$$

$$R' = \nu I - \mu R - \gamma R + \omega R_V$$

$$V' = \epsilon p\mu - \mu V - \beta_V(t)V(I + I_V) + \gamma_V R_V - \omega V$$

$$I'_V = \beta_V(t)V(I + I_V) - \nu_V I_V - \mu I_V - \omega I_V$$

$$R'_V = \nu_V I_V - \mu R_V - \gamma_V R_V - \omega R_V,$$

with $\beta(t) = b_0(1 + \bar{b}\cos(2\pi t + \phi))$ and $\beta_V(t) = (1 - \alpha)\beta(t)$, for $0 \leq \alpha \leq 1$, where α represents... . (We may relax the lower bound on α later.) The model is illustrated in Figure 4.1. *I'm confused about the rationale for relaxing the lower bound on α? '

These were Alexandra's notes on the first part of my section introducing the model. Her first edit added a motivation, which is the kind of 'duh' moment that had me slapping my forehead (because of course we need a reason to be extending this model!). She changed $\beta(t)$ from a parameter to a function (which it is, because of the dependence on time) and suggested adding the word 'protectiveness' to the waning of the vaccine. After introducing the model, she asked what α represented biologically and also questioned why I was allowing the lower bound to relax.

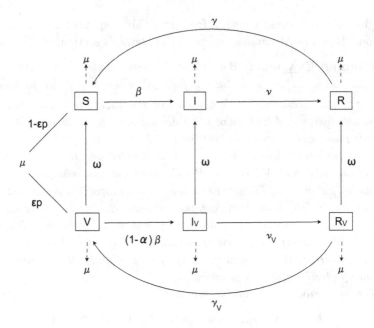

Fig. 4.1. The RSV model.

All of these are sensible additions and questions. In several cases, the answers were perfectly clear in my head and sometimes elsewhere in the manuscript, but they needed to be written down here. It was very clear to me what α represented, but it obviously wasn't to others. And I had written the parenthetical about relaxing the lower bound on α because I knew I'd need to do that in order to get the infection spikes, but Alexandra (and every other reader) was seeing this for the first time here, so of course it was going to need a justification.

Next, Alexandra looked through the initial analysis and had a few more comments and additions about the mathematical logic.

WORKED EXAMPLE: *'From $c_1 = 0$, we find* I am confused about this step here, sorry! How did we get from setting $c_1 = 0$ to determining R_0?

$$R_0 = \frac{\beta \bar{S}(\nu_V + \mu + \omega) + \beta_V \bar{V}(\mu + \nu + \omega)}{(\mu + \nu)(\mu + \nu_V + \omega)}$$

(This is equivalent to the value found using the next-generation method.)

This is a pretty minor note, and it turns out it's a very standard step... but if she's confused by it, then it still needs to be clarified! Finally, she also added a penultimate paragraph to the Discussion, because there were further implications that she'd identified.

WORKED EXAMPLE: *'Since we extended the model of Weber et al. (39), our model inherited many of the assumptions from that model, such as mass-action transmission, a constant birth rate and that the birth and death rates were matched, resulting in a constant population.*

In our model we considered RSV transmission dynamics for a single age class, in order to allow for the model to be analytically tractable. Given we were examining the broad population-level impacts in a large population, we considered this a reasonable model simplification. Further, it has been shown that for a similar compartmental RSV model, including multiple age classes did not change the bifurcation structure of the model (16). However, different vaccine candidates for RSV are being developed for distinct key age groups – infants, young children, pregnant women, and the elderly (30). This means that future models that explore the specific implications of vaccines for these target groups may need to incorporate additional age classes.

A vaccine that targets RSV infection has the potential to significantly reduce the overall prevalence of the disease, but...'

This is excellent. It discusses a further limitation but also ties it back to the biology (which she knew much better than I did) and talks about potential future directions. All of these additions require me to rethink aspects of the manuscript and to consider what needs to be changed or thrown away. A good beta reader is like gold.

BAD EXAMPLE : *'Malaria is a disease that has existed for centuries dating back to 2700 BCE (Centers for Disease Control and Prevention [CDC], 2012a; Cox, 2010). This disease has lasted over these many years and since then many discoveries have been made on the illness. The first antimalarial was discovered around the 17th century when Spanish missionaries had found a type of tree bark being used among aboriginals as a treatment of malarial fever. The medicinal ingredient among it, quinine, became one of the first medications sought against the disease (CDC, 2012a). With time, in 1880, a discovery was made by Charles Louis Alphonse Laveran by finding the Plasmodium parasite within a malaria patient's blood (Cox, 2010). Greater research by Camillo Golgi found the differing forms of the parasite along with the replication process of the disease through merozoites (CDC, 2012a).*

Along with the discovery of the malaria parasite Plasmodium, Ronald Ross would find in 1897 how malaria was being transmitted through infection by the mosquito vector (CDC, 2012a; Cox, 2010). Laveran had originally hypothesized that the disease was spread by mosquitoes, and through tests among birds, Ross demonstrated that mosquitoes would become infected after taking blood meals on infected birds and then pass the Plasmodium parasite to a susceptible bird upon taking further blood meals (CDC, 2012b). Ross, Lavaran, and Golgi would each eventually go on to earn Nobel prizes for their respective works in the greater understanding of malaria (CDC, 2012a). These achievements along with many others from those involved in research and understanding of the disease and the parasite have led to today's knowledge on the transmission processes and methods to combat the disease (Cox, 2010).

From the knowledge gained on this disease and how it is caused, preventative measures have been sought and are currently being used today around the world to prevent transmission of this deadly illness.'

This was the opening from a student thesis I was supervising. The student was a good scientist but not a strong writer. In particular, he had a tendency to waffle and to write sentences that superficially seemed reasonable but that seemed to lose their way.

The second sentence of the first paragraph is particularly weak. Why do we need 'these many years'? The fact that 'many discoveries' have been made is true but redundant. And the final paragraph says nothing at all. I mean, sure, those facts are all true... but they're coming at us from 30,000 feet, when what we need are specifics.

I asked the student to streamline it himself, so he made a second attempt. The first paragraph wasn't changed, but the third was deleted altogether, and the second was revised. I've bolded the changes.

BAD EXAMPLE : *'Along with the discovery of the malaria parasite Plasmodium, Ronald Ross would* **discover** *in 1897* **that mosquitoes could transmit the infection** *(CDC, 2012a; Cox, 2010). Laveran* **and his colleague Patrick Manson were one of a few that originally** *hypothesized the disease was spread by mosquitoes and, through tests among birds, Ross demonstrated that mosquitoes would become infected after taking blood meals on infected birds and then pass the Plasmodium parasite to susceptible birds upon taking further blood meals (CDC, 2012b; Cox, 2010). Ross* **and Lavaran** *would each eventually go on to earn Nobel prizes* *for their respective works in the greater understanding of malaria (CDC, 2012a). These achievements, along with many others from those involved in research and understanding of the disease and the parasite, have led to today's knowledge on the transmission processes and methods to combat the disease (Cox, 2010).'*

There are a few changes here, but they're mostly factual corrections. Golgi did win a Nobel prize, but for his studies of the structure of the nervous system, not malaria. However, some of the changes introduce even more awkwardness: 'Laveran and [...] Manson were one of a few' just doesn't make grammatical sense (since they were two people, not one). The construction 'Ronald Ross would discover' is a weak one. Why not just 'Ronald Ross discovered'? Let's see what happens after a third attempt, which happened as a result of my notes.

BAD EXAMPLE : *'Malaria is a disease that has existed for centuries, dating back to 2700 BC (Centers for Disease Control and Prevention [CDC], 2012a; Cox, 2010). The first antimalarial was discovered around the 17th century,* **where a type of tree bark was used** *as a treatment of malarial fever. The medicinal ingredient found within it, quinine, became one of the first medications against the disease (CDC, 2012a). In 1880, Charles Louis Alphonse Laveran found the Plasmodium parasite within a malaria patient's blood (Cox, 2010).* **Further** *research by Camillo Golgi found differing forms of the parasite, along with the replication process of the disease through merozoites (CDC, 2012a).*

In 1897, Ronald Ross **discovered** *that mosquitoes could transmit the infection (CDC, 2012a; Cox, 2010). Laveran and his colleague Patrick Manson* **had** *originally hypothesized the disease was spread by mosquitoes and, through* **examining** *birds, Ross demonstrated that mosquitoes would become infected after taking blood meals* **from** *infected birds and then pass the Plasmodium parasite to susceptible birds upon taking further blood meals (CDC, 2012b; Cox, 2010). Ross and Lavaran* **earned** *Nobel prizes for their respective works in the greater understanding of malaria (CDC, 2012a).'*

This is much more streamlined. In the first paragraph, the second sentence has been removed, while the third (on antimalarials) has been reduced to a much more factual one that also avoids the word 'aboriginals', which has difficult connotations. Irrelevant constructions like 'With time' have been removed, while 'Greater research' has become 'Further research', which is a lot more accurate.

The second paragraph has changed a lot. 'Ronald Ross discovered' is the much more sensible choice. Instead of 'through tests among birds', we now have 'through examining birds', which is a lot nicer. Even some of the smaller changes, like taking a blood meal 'from' a bird, rather than 'on' a bird do a lot of work to make the writing more logically sound. Never underestimate the power of simplicity in the right moment, as we saw when Laveran and Manson 'had' originally hypothesized, rather than being 'one of a few that' originally hypothesized. Ross and Laveran now simply earned Nobel prizes, rather that 'would eventually go on to earn', because we don't need to know that there was a passage of time here. The final sentence has also been excised, because it was too woolly.

This still isn't perfect, of course. The spelling of Laveran/Lavaran is all over the map, and there are constructions that could definitely be improved. But it reads a lot better after several rounds of editing. There's no way that first draft was publishable. Sometimes, you need strong editing in order to save you from yourself.

4.2   Feedback

As we saw from Alexandra's notes, beta readers are like gold. Don't waste this amazing opportunity for improvement. What that means is you have to deal with the issues they raise. Thank them profoundly for their time and offer to repay them in kind, because you can be just as valuable a beta reader for them as they are for you.

Okay, what to do once you have this valuable feedback? Answer: do something. Not necessarily the thing they suggested, but something. It's my experience that most people are really good at detecting crap, but not too good at actually dealing with it. A lot of people will yell at the TV because of holes in the logic of the show they're watching, although that doesn't make them good scriptwriters. But it also doesn't mean they're wrong.

If someone raises something, it needs to be addressed. You don't have to implement their solution, but you do have to find one. And your solution needs to be at least as good, if not better, than theirs. Sounds challenging? This is why you're the expert. As a writer, your job is problem-solving.

The good news for all scientists reading this is that the same skills that your training gave you will work just as well on the logical flow of sentences and the rigour of paragraph structure as they do in a scientific investigation. Scientists should all be competent writers, by virtue of who they are. The fact that they're not is, I think, simply because no one has taught them how. So harness your superpowers.

As an aside, the reason I became a decent writer is because I was reviewing novels on the internet, just for fun. People noticed that my reviews were unlike anyone else's, because all the English majors were looking for turns of phrases and metaphor and beautiful imagery, whereas I was focusing on the strict logical structure of the books... and the authors were generally good at writing beautiful prose, but they weren't always so crash-hot on logic and rigour. My reviews started to make a splash, including among the authors themselves, because I was dissecting the plots with a scalpel. It took me a while to realise that this was unusual, because I was simply applying the same logical thinking to the plot of a novel as I would to the structure of a mathematical proof. I hadn't realised it, but mathematics had given me a superweapon. So if science or math is your background, leverage that into becoming at the very least a logical and consistent writer, because you already have these skills.

Weaknesses need to be identified and strengthened. This means weaknesses in both the logic and the writing. You need to be detail-oriented enough to spot the flaws in the logical construction of a sentence, as in the malaria example in the last section. But you also need to pull back and look at the big picture and see how the overall manuscript flows and discover what's missing.

Do this before submitting the manuscript for peer review. The reviewers will likely spot these problems if you don't... and even if they miss them, the readers will likely see them, and you'll become a laughing stock. Remember that your reputation is critical in academia, and your published papers are the major way you'll be judged.

Officially, you submit the manuscript to an editor, who then decides whether or not to send it out for peer review. Once you submit, there are three outcomes. The first is that it gets accepted with no changes. (This almost never happens.) The second is that it'll be rejected outright. (This happens a lot. Sometimes the editor rejects it outright, often because it's a bad fit for the journal. Sometimes it goes for peer review and gets rejected by the reviewers.) The third is that it will come back to you for revisions, possibly extensive revisions. (If this happens, rejoice: you're over the hump!) The aim of this book is to move your paper from the second category to the third.

The biggest danger is the second category: immediate rejection, with no chance of resubmission — and this is where the quality of the writing can really make a difference. If the reviewers have to spend too much energy nitpicking your writing and your logic, they'll reject your manuscript. Even if your work is good, you don't want to waste their time and energy. Generally, I've found that once I get over the hump of a revise-and-resubmit, into the third category, the manuscript is very likely to eventually get accepted (even if there's often some serious hard work to be done before that happens). However, even if you do manage to get past that initial hump, you'll have so much to do that your manuscript is going to become a sort of Frankenstein creature, with bits added and altered because of the reviewers. That's going to happen to some degree anyway, but the more you can master the art of revision, the easier it is.

As mentioned in the Preface, being a decent editor has made me a much better writer. I'd say I'm a competent writer, but I'm a brilliant editor. So that means that, as a writer, I simply write and trust that someday an amazing editor will come along and fix all my mistakes. That person happens to be me, but it doesn't have to be if you can find a good editor elsewhere. And by not second-guessing myself as a writer, it means I don't get writers' block.[2] I cannot recommend learning editing skills highly enough.

What this has meant for me is that certain doors open more quickly than they do for other people. Because I'm so used to revise-and-resubmit from academia, I know exactly what to do when someone criticises my work. I've had non-academic books moved up in the publishing schedule because the other writers were too frozen from having to implement edits, whereas I just went ahead and incorporated them, much to my editors' delight. It also means I've handed publishers near-impeccable manuscripts, which is a godsend to busy publishing houses. These are all skills that you can learn.

The key to everything is to remember who you're writing this for. You're not writing it for the editor or the publisher. You're not writing it for yourself or your friends. No, you're writing this for the *reader*: that ineffable general reader who wants to hear what you have to say and is willing to give some of their valuable time to do so. Your entire goal in the act of writing is to make life easy for the reader, so that the prose runs smoothly, the logic is impeccable and they don't pause, flip back several pages and say

---

[2] I wrote the first draft of this entire chapter that you're reading in a single evening, for instance. That's because the first draft has the lowest stakes of all the writing that will happen.

'Wait, that can't be right...' because they've caught you in a contradiction or simply been confused by your poor writing.

Stumbling over gaps in logic or typos or poor constructions are like catching sight of a boom mike in a TV show: they take you right out of the action at best, and make your work laughable at worst. You don't want people making fun of your paper any more than a director wants people laughing at the screen.

Next, I'll give my responses to Alexandra's comments from the previous draft (see Page 62), showing how to bolster every point. Her comments are still highlighted; my additions are underlined.

WORKED EXAMPLE: ' *More recent studies conducted for the setting of rural Kenya have focussed on the likely benefits of vaccination for particular target groups (22,35).* *To the best of our knowledge, there are no theoretical models that examine the impact of an RSV vaccine analytically.*

Here, we examine the effects of a prophylactic vaccine on the transmission of RSV *in a single age class* .'

This addresses the missing segue that Alexandra mentioned. It basically just formalises exactly what she said. But what about her suggestion (on Page 61) to remove the paragraph about the newborns? To address this, I rewrote the Methods section (Page 63):

WORKED EXAMPLE: '*We extend the SEIRS compartmental model for a single age cohort described by Weber et al. (39)* *to include a vaccine strategy for RSV where a fixed proportion of newborns are vaccinated before infection. (This is equivalent to the situation where pregnant mothers are vaccinated before giving birth.)* *We assume that the leaving rate μ is unchanged across all classes and that there is no disease-specific death rate. We scale the entry and leaving rates so that the population is constant.*'

I changed 'basic' to 'SEIRS', which was just a tweak for clarity, but the most important change here is to note that vaccination occurs a) before infection and b) to explicitly point out that the vaccine is being given to pregnant women. That's a very minor change... but it entirely justifies the newborns paragraph in the Introduction. This is the perfect solution to the problem Alexandra raised, even though it's not at all her suggested solution of cutting the paragraph.

It's now time to tackle Alexandra's issue with what α represents, from Page 63:

WORKED EXAMPLE: *'with $\beta(t) = b_0(1 + \bar{b}\cos(2\pi t + \phi))$ and $\beta_V(t) = (1 - \alpha)\beta(t)$, for $0 \leq \alpha \leq 1$,* where α represents *the efficacy of vaccination in preventing infection. (We will relax the lower bound on α later in order to examine some theoretical scenarios.) The model is illustrated in Figure 4.1.'*

Here I've addressed the issue of explaining what α represents biologically, as well as explaining why we might want to relax the bound on alpha. I deliberately left this general, as I didn't want to give away the punchline at this stage, but it's enough to justify the comment that Alexandra made. Next up is the gap in logic that led from c_1 to R_0 from Page 64:

WORKED EXAMPLE: *'We use the method of the constant term of the characteristic polynomial to determine the reproduction number (19). Rearranging $c_1 = 0$, we find*

$$R_0 = \frac{\beta \bar{S}(\nu_V + \mu + \omega) + \beta_V \bar{V}(\mu + \nu + \omega)}{(\mu + \nu)(\mu + \nu_V + \omega)}$$

(This is equivalent to the value found using the next-generation method.)'

This issue is solved with a single line and a reference. That's because, as mentioned above, it's a fairly standard step... but it still needs to be strengthened. And don't underestimate the power of a reference to justify something.

BAD EXAMPLE : *'A technique that can be used to review malaria transmission and intervention strategies are through ordinary differential equations and impulsive differential equations as demonstrated (1). The investigators' model depicted varying factors in the modelling of indoor residual spraying's reduction on malaria transmission (1). Effective spraying periods for insecticide were assessed using their mathematical model based on level of effectiveness of the IRS and parameters affecting the impulsive differential equations, leading to an understanding of spraying intervals of both scheduled and unscheduled spraying times (1). An understanding of IRS effectiveness can be done through modelling ordinary and impulsive differential equations and is further discussed in the methodology of this thesis.'*

Here's a paragraph from the middle of the same malaria thesis as in the previous section. The first sentence wanders about far more than it should, so that by the time you get to the end of it, you've forgotten how it began. The second sentence is probably redundant.

BAD EXAMPLE : 'Ordinary differential equations and impulsive differential equations can be used to model malaria transmission and intervention strategies *as demonstrated* by *(1)*. *Effective spraying periods for insecticide were assessed using their mathematical model based on level of efficacy of the IRS and parameters affecting the set of impulsive differential equations, leading to an understanding of spraying intervals for both scheduled and unscheduled spraying times (1). An understanding of IRS efficacy can be done through modelling ordinary and impulsive differential equations and is further discussed in the Methodology.'*

Version 2 is the student's attempt to tighten things up, to only limited success. For one thing, the 'as demonstrated by' is redundant and could be simply replaced by the reference. Constructions such as 'can be done' are very weak. Another problem, which is a bugbear of mine, is having the same reference appear multiple times in a row. It's not wrong, but it's clunky. Why not have just one at the end? (Answer: because then the earlier ones look like they're missing references, which isn't acceptable.) Better to splice in some alternating references.

Side note to say that different publishers will use different referencing styles. Some use numbers in brackets, like I do in this book. Others use author-year or superscripts. And the citation order may vary: sometimes they're in alphabetical order, other times in chronological order. You'll need to be adept at rolling with the changes, because submitting the same manuscript to different publishers may requiring changing the reference style and order multiple times. If you use LaTeX (and I recommend that you do!), then all these issues are easily solved with one line of code.

We also need more details about what ordinary and impulsive differential equations actually are. (Note: this sort of baseline definition isn't required in a paper, but it is needed in a thesis. A thesis is expected to have a lot more background, but it's important not to waffle; the explanations should be there, but they need to be tight.) My changes are underlined.

BAD EXAMPLE : *'Ordinary differential equations and impulsive differential equations can be used to model malaria transmission and intervention strategies. Ordinary differential equation models calculate the differing states of populations, such as humans that are susceptible to an illness or infected with a disease, and the rate of change of these states over time using derivatives (1). Impulses can be added to the ordinary differential equations at specified points in time to approximate an instantaneous increase or* *decrease in a population based on, for example, a reaction to an intervention* *such as IRS (2). These ordinary differential equations combined with* *impulses creates a mathematical model of impulsive differential equations* *to understand the behaviour of a disease like malaria (1). Smith? and* *Hove-Musekwa (2) demonstrated these methods through examining effective* spraying periods for insecticide using their mathematical model based on the level of efficacy of IRS and parameters affecting their set of impulsive differential equations. *The investigators' results led* to understanding spraying intervals for both scheduled and unscheduled spraying times (2). An understanding *of LLIN and* IRS efficacy can be achieved through ordinary and impulsive differential equation models. *This is further discussed in the Methodology.'*

Note how succinctly the second and third sentences here define what are actually very complex topics. No two consecutive references are identical. A more sophisticated version would mix and match multiple references, rather than toggling back and forth between just two, but this is okay for what it is.

Once again, this isn't perfect, but it's come a long way from where it started. The clunky, torturous sentences are mostly gone. Sentences are succinct without being too short. Big issues, like missing information, have been solved.

Slowly but surely, the paragraphs are rising around the house. This may take several revisions, as we've seen, or it may take external input, but it's crucial to (a) find the issues and (b) resolve them. As we do that with every paragraph, the manuscript gradually gets better and better. This process takes time, and it takes investment of energy, but it's the key step in the entire writing process. After all, at the end of the day, all writing is editing.

Okay, so now it's done. You've written a manuscript! More importantly, you've self-edited and revised a manuscript, and you've run it by trusted beta readers. So now it's ready to send off for professional assessment.

We'll talk about what to do when the reviewers respond in a later chapter, but first I'd like to congratulate you for getting this far. One of my colleagues always celebrates when she gets to this stage, because her reasoning is that this is where you've done everything you can do. Everything else is in the hands of the gods, but this is the point that's (almost) entirely your achievement. So relax and enjoy while you wait for the reviewers' comments. Because, well, you'll need to be refreshed for when that happens.

While we wait for the reviewers, let's take a brief aside to talk about micro-editing, at the sentence level.

Chapter 5

Sentence Logic: Following Through What You Started

5.1 A brief interlude about following your own ground rules and not prevaricating all over the place or using awkward things sometimes or trailing off...

5.1.1 *Where does your article begin?*

I've stated a few times that the unit of writing is the paragraph. That's absolutely true, and hopefully I've convinced you of that when designing the structure. But there's also the matter of sentence structure, which matters quite a lot.

Writing is mix of big ideas and granular details. Your novel needs to have a plot, but it also needs to have beautiful descriptions and elegant metaphors. A plot-heavy book that has no poetry can be a page-turner, but it won't win any literary prizes. And a novel full of beautiful writing that takes you nowhere isn't going to set the world on fire.

An academic paper doesn't have to be quite so poetic. However, it still needs to satisfy both basic tenets: does it tell a good story, and does it tell it well? Most people tend to be good at either one or the other, but not both. I know I had to learn the big-picture stuff, because I was quite strong on individual sentences, but not so good at the overall direction. One tactic I used for a long time was to find co-authors who were good at the big picture but who needed a decent editor to make their words sound like English. That kind of collaboration can bolster your weaknesses in a really productive way.

However, it's also possible to learn the skills yourself. I spent some time studying what my co-author did when editing our stuff. Sometimes he'd consign whole paragraphs to the bin. Or he'd swap sections over entirely. One of the best lessons I learned was 'Where does this piece begin?' It was rarely where I thought it did. Mastering that was a huge leap in my writing and editing abilities.

Interruption: I had initially just presented the above paragraph and moved on to the bulk of this chapter, which is about sentence logic. However, Anthony suggested that we needed an example of knowing where to begin, because it was something lots of people struggle with, him included. Take a look at the next example, which was an introduction written by one of my students, and see if you can work out where it really begins. (Hint: break it down into the skeletal structure.)

WORKED EXAMPLE: *'An Ebola virus outbreak results in negative impacts on affected communities. One of the most significant impacts is the sector-wide weakening of human resources, as observed in the three epicentre countries in the 2014 outbreak, as a result of increased mortality rates. More specifically, the Ebola outbreak in 2013–2014 negatively impacted affected communities' education sector. Schools experienced persistent losses of teachers, education staff and students due to the outbreak resulting in many of them being understaffed. Furthermore, the fear of spreading the virus at school resulted in widespread school closures, restricting access to a formal education. In Sierra Leone, Guinea, and Liberia 1848 schools closed between June 2014 and January 2015. This prevented many children from attending school during and after the outbreak.*

Since the Ebola virus took many adult lives during the epidemic, children experienced the additional loss of one or both parents. This resulted in many children being orphaned. Although orphans are usually taken care of by next of kin in West Africa, in some cases the stigma and fear surrounding the contraction of the Ebola virus proved to be stronger than family ties, leaving orphaned children abandoned. Children, particularly orphaned children, became more vulnerable during the epidemic as the state of health in a country deteriorated. Moreover, during the Ebola virus outbreak in West Africa, there was an observed increase in children dying from other vaccine-preventable diseases, such as measles, exacerbating existing poor health outcomes among children affect by the outbreak.

The health sector itself was also impacted and experienced weakening of availability, accessibility and quality of health services due to the loss of human capital. With most healthcare resources and personnel allocated to efforts aimed at controlling the outbreak, many found themselves unable to obtain their regular healthcare treatment and services. This was compounded further when some clinics were closed permanently due to the fear of contracting the Ebola virus experienced by many healthcare practitioners.

This resulted in the weakening of the affected community's health system, and made access to specific services, specialist care and primary care more difficult. Lastly, the health systems in West African countries affected by the recent epidemic were unprepared for an Ebola outbreak. From the beginning of the outbreak, communities in these countries lacked sufficient resources needed to contain the outbreak while at the same time continuing to respond to the demand for on-going health care.

The Ebola virus further affected countries where a significant proportion of their population was already living in poverty. The outbreak in Guinea, Liberia, and Sierra Leone contributed to a reduction in the production of goods and services in multiple sectors, reversing economic growth in these countries. Neighbouring countries such as Senegal, Cote d'Ivoire, Guinea Bissau, and Mali also experienced severe consequences due to the impact of the epidemic on respective poverty levels. The net increase in poverty and net decrease in production of food observed in affected countries exacerbated existing levels of food insecurity. In the three epicenter countries of the epidemic an increase in undernourishment rates was observed, highlighting the large proportion of individuals considered to be food insecure.

Within this context, the transmission of the Ebola virus to humans is known to occur through direct contact in three ways: from a reservoir, between humans, and from the infected deceased. Fruit bats, which experience asymptomatic infections of the virus, serve as a reservoir of the Ebola virus and contribute to its transmission to other wild animals and humans. In West Africa, wild animals such as fruit bats are hunted, sold, and consumed creating the opportunity for the exchange of bodily fluids resulting in zoonotic transmission of the virus. The virus is spread from human to human through direct contact of bodily fluids such as blood, saliva, vomit, feces, urine, sweat, nasal secretions, semen and genital secretions. Once infected, with an incubation period averaging 11 days, individuals begin to experience an onset of symptoms including: headaches, vomiting, loss of appetite, diarrhea, stomach pains, lethargy, aching muscles or joints, difficulties swallowing and breathing in addition to unexplained bleeding. Following this, symptoms further progress to weaken liver and kidney functions in addition to more serious hemorrhagic symptoms such as internal and external bleeding. Upon developing these symptoms, infected individuals are able to spread the virus to others, making them infectious.'

I'll first note that nothing here is particularly egregious. It's reasonably well written and interesting. But a lot of it is simply 'true facts': things that are certainly true, but what are they contributing to the narrative?

The answer is to entirely cut the first four paragraphs, along with the words 'Within this context'. Yes, for real. You don't need them. That last paragraph tells us everything we need to know, without bogging us down in anything else. Trimming the fat is a crucial skill in editing.

5.1.2   *Sentence logic*

This chapter deals with the granular side of things, in particular the issue of sentence logic. We've seen some of that in the previous chapter, but I think it's worth spending some time on, because most bad writing is actually an inability to follow through on the structure of a sentence.

Specifically, you need to follow the logic of each individual sentence. Small errors here can quickly make a sentence incredibly clunky. It's a bit like the way you can spot a non-native speaker of a language when they make a single tiny error... that no one who grew up speaking it would make.[1] When it comes to writing, these kinds of errors pop up all the time, including from people for whom English is their first language.

Your academic writing doesn't have to be Shakespeare, and it doesn't have to have elegant metaphors, but it does need to make basic sense. Let's work through some examples.

BAD EXAMPLE : *'Symptoms of RSV range from those of a cold, more severe afflictions such as bronchiolitis and pneumonia.'*

You might recognise this as one of my own sentences, from Page 38 and flagged on Page 40. It starts going somewhere and then doesn't get there. Let's see if we can fix it.

(RE)WORKED EXAMPLE: *Symptoms of RSV range from those of a cold to more severe afflictions such as bronchiolitis and pneumonia.*

Swapping a comma for the word 'to' is a tiny change, but it makes a massive difference.

BAD EXAMPLE : *'This is where the recent studies on how to provide insights into the effects of these spatial aspects on the overall mosquito and human infection dynamic.'*

Hopefully by this point you can read that and are able to see what's wrong with it. Or at least know that there is something wrong. Let's simplify

---

[1]I've definitely made that kind of error in French. You gender a table wrong, you're going to get caught!

the words a bit while keeping the overall structure: 'This is where studies on how to provide insights...' So far so good. It sets up an expectation that those studies about providing insights will actually be doing something by the time the sentence is completed. But they don't. The sentence starts out heading in a direction that never pays off. And because it's a long sentence, it's tough to remember that there's supposed to be a capstone. Instead, it just sits oddly in the brain.

The real problem is that the sentence is using 'how' as a verb, so everything is inadvertantly pivoting around that. It looks like it should say something like 'This is where studies on how to provide insights [into these effects] will pay off' and has forgotten the last bit. Instead, the issue here is that 'how' is the wrong pivot, so the simple answer is just to remove 'on how to'.

(RE)WORKED EXAMPLE: *'This is where the recent studies <u>provide</u> insights into the effects of these spatial aspects on the overall mosquito and human infection dynamic.'*

That's much stronger, because the pivot is 'provide', not 'how'. But it's not easy to see that from the outset.

BAD EXAMPLE : *'The COVID-19 pandemic lasted for almost three years since the outbreak started in March 2020 in Canada.'*

This was the first line in an Abstract written in December 2022. The first thing to note is that the COVID-19 pandemic was still going, but the wording implies that the pandemic had ended after three years. (If only!) Second, the 'in Canada' at the end is awkwardly tacked on and would be better if it were integrated, since this is a manuscript that's about the Canadian pandemic. So an improved version would be 'The COVID-19 pandemic in Canada has lasted for almost three years since the outbreak started in March 2020.' However... why do we even need this sentence? The pandemic has lasted for almost three years worldwide, so what's special about Canada here? Answer: nothing. Delete the sentence altogether.

BAD EXAMPLE : *'The hospitalizations and deaths have declined from the first waves, along with the nonpharmaceutical interventions and the vaccination campaign starting in December 2020, despite keeping emergence of highly contagious variants.'*

The first part is mostly fine. Hospitalisations and deaths have declined, yes. (Although no leading 'The'.) But have they declined from the first waves or have they declined *since* the first waves?

Next, we run into an inadvertently hilarious statement: apparently, the nonpharmaceutical interventions and vaccination have also declined from the first waves. Which is quite a feat, since there was no vaccine during the first wave! (Presumably there's negative vaccination by now?)

Finally, what on earth does 'keeping emergence' mean? Reading this sentence, I feel like I know what it wants to say, but it's actively working against my interpretation at every turn. Here's an improved version:

(RE)WORKED EXAMPLE: *'Hospitalizations and deaths have declined since the first waves, thanks to nonpharmaceutical interventions and the vaccination campaign, which started in December 2020, despite the emergence of highly contagious variants.'*

I think that says the same thing, but without the words actively fighting the reader.

BAD EXAMPLE : *'We conducted the counterfactual scenarios of no mask mandates, no vaccination and one month earlier of these two strategies and compared the number of cases, hospitalization and deaths with that reported under the actual situation.'*

Is 'conducted' really the best choice here? Does one conduct a scenario? You can conduct a survey or an orchestra, but I would think you *examine* a scenario. Also, it's minor, but the word 'of' isn't great here either. It's too informal; it would be significantly improved with a colon.

The 'one month earlier' part gets messy, because I don't think it's clear what's actually happening here. (It's another counterfactual, testing to see what would happen if the mandates had been introduced a month earlier than they were in reality, but I think that's in danger of being lost.) What is actually happening with the one-month-earlier strategies? There's a verb missing here.

The use of the two 'and's to join things is really stretching the sentence to breaking point, because these 'and's are doing wildly different things. Much easier to break this into two sentences. And do you compare to or compare with? Let's ask Shakespeare: 'Shall I compare thee *with* a summer's day?' Urk. I don't think so. Finally, the phrase 'under the actual situation' is an odd way to describe reality.

(RE)WORKED EXAMPLE: *'We examined several counterfactual scenarios: no mask mandates and no vaccination, as well as instigating these strategies one month earlier.'*

This is much tighter, because it clumps the three options together as clear counterfactuals. Scenarios are examined, not conducted; strategies are actually instigated.

(RE)WORKED EXAMPLE: *'We compared the number of cases, hospitalization and deaths to the reported numbers.'*

Breaking the sentence helps streamline things. Cases are compared to the reported numbers. It's clean and straightforward. Sometimes the simplest solutions are the best.

BAD EXAMPLE : *'We found that mask mandates reduced 35% of cases and 69% of death, among which the largest number of cases, hospitalizations, and deaths were averted in the 60+ age group.'*

Did they really reduce 35% of cases or did they reduce cases by 35%? And what does 'mandates reduced [...] 69% of death' actually mean? Yes, the reader can figure it out, but they shouldn't have to.

(RE)WORKED EXAMPLE: *'We found that mask mandates reduced cases by 35% and deaths by 69%, among which the largest number of cases, hospitalizations and deaths were averted in the 60+ age group.'*

It's a small change, but it makes a big difference for clarity.

BAD EXAMPLE : *'The vaccination program contributed significantly to the mitigation of the third wave of COVID-19 outbreak in Ontario, reducing 92% of symptomatic cases, while the largest cases, hospitalizations, and deaths averted were in 20–59, 60+ age groups.'*

Again, it would be better as 'reducing symptomatic cases **by** 92%'. But let's look at the second part here: the largest reductions were in the 20–59 and 60+ age groups... so, basically, everyone? That kind of makes the word 'largest' useless here. At this point, the age ranges aren't all that helpful. One could perhaps say that reductions were larger for adults than children. Or simply delete, as I don't think this part is adding anything.

BAD EXAMPLE : *'On the other hand, a one-month earlier vaccination program will significantly decrease the number of cases and hospitalizations, while one-month earlier mask mandates will not.'*

Let's start with the hyphens. You use a hyphen in this context for a noun-noun compound adjective that's describing something. Is 'one-month' describing 'earlier'? It is not. It should be 'a one-month-earlier vaccination program'.

My favourite word in this sentence is 'will'. Let's recap: under a counterfactual scenario, if we had theoretically introduced the mandates a month earlier, back in 2020... these will make a difference in the future. Uh huh? Will they really? Something that is physically impossible to do **will** change the future? I don't think so.

This is why word choice is so important. I read that sentence and spent the next five minutes thinking deeply not about the results the authors wanted to convey but about how time travel works and whether a hypothetical action in the past can have a definitive outcome in the future.

(RE)WORKED EXAMPLE: '*A one-month-earlier vaccination program could have significantly decreased the number of cases and hospitalizations, but one-month-earlier mask mandates would not have.*'

This moves the sentence firmly into the realm of the hypothetical.

BAD EXAMPLE : '*Our model demonstrates that mask mandates play a vital role in saving lives in the first wave of the COVID-19 outbreak and the vaccination program was crucial to averting the cases and hospitalizations after it was implemented.*'

This is pretty good, except that the tenses are all over the map. The counterfactual was in the past, so the tense should be as well. It needs a few other tweaks as well.

(RE)WORKED EXAMPLE: '*Our model demonstrates that mask mandates played a vital role in saving lives in the first wave of the COVID-19 outbreak and that the vaccination program was crucial to averting subsequent cases and hospitalizations after it was implemented.*'

The insertion of the word 'that' makes this stronger (by providing a more direct link to the first part of the sentence) and the word 'subsequent' is much stronger than 'the cases' while also reinforcing the timeline.

BAD EXAMPLE : '*Our model assume host population(human) and a vector population (animal). Host population is subdivided into four compartments Susceptible S_H, Asymptomatic A_H, Infective I_H and Recovered R_H.*'

'Our model *assumes*' is correct grammar. But that's not really the problem here. While models often make assumptions, the fact that humans and animals are involved in this disease (leptospirosis) isn't one of them; it's a fact. The four compartments need a colon. Human and animal shouldn't be singular here.

(RE)WORKED EXAMPLE: *'Our model <u>considers both a</u> host population <u>(humans)</u> and a vector population <u>(animals)</u>. <u>The</u> host population is subdivided into four compartments<u>:</u> Susceptible S_H, Asymptomatic A_H, Infective I_H and Recovered R_H.'*

BAD EXAMPLE : *'The state of leptospirosis dynamics described above will be studied in domain Ω_H, Ω_H can be described by a vector with non-negative components subset Γ, given by;*
$\Gamma = (S_H, A_H, I_H, R_H, S_A, E_A, I_A) \in R^7$, differentiable in the bounded domain $\Omega_H \in R^7$ for a given time $t \in R^7$.'

The first comma is a comma splice, which is too weak to join clauses. It should be replaced by a semicolon at the very minimum or else a joining word like 'and'. But that's the least of the problems here. The domain is labelled Ω_H, but H is a subscript for humans, whereas the state variables (the ones that we differentiate) include both H and A. So Ω really needs to have no subscript.

The description of what Ω_H is would be more elegantly written in words. It's simply the non-negative values. Also: why do we need to specify that? Answer: so that the model is well-posed. That is, solutions that start non-negative will remain non-negative, so that you don't get unrealistic things like negative people. (I for one always try to avoid negative people...)

(RE)WORKED EXAMPLE: *'The model dynamics will be studied in the domain Ω, consisting of the nonnegative segment of $\mathbb{R}^7$. This ensures the model is biologically well posed, since all parameters of our model are positive.'*

5.2   Homework

Hopefully by this point you've seen enough examples to at least know when something is wrong. Even better is if you can fix them. Here are a few examples for you to try yourself.

BAD EXAMPLE : *'The quest for a world without leptospirosis is achievable if the control and the eradication of leptospirosis is taken serious owing to it strong biological and socio economical importance.'*

My guess is that you probably stumbled at the word 'serious' and had difficulty finishing the sentence after that. Try rewriting this sentence in better English.

BAD EXAMPLE : *'The Tonado plot and the sensitivity analysis of our model revealed factors that mostly influence our model and in fact the real life as leptospirosis is concerned.'*

'Tonado' should be 'Tornado' (a type of plot). Rework this sentence into readable English. Don't be afraid to move things around; the changes don't have to be tweaks.

BAD EXAMPLE : *'Among all these parameters, our results strongly point vaccines efficacy as the most important parameter that can reduce the transmission of leptospirosis as any little increase in the vaccine efficacy will greatly reduce the basic reproduction number of our model and by implication reduction in the transmission process of leptospirosis.'*

This time, rewrite the paragraph entirely in your own words, while keeping the meaning intact. When you're done, give your paragraph another edit to make it even stronger.

This chapter has been a bit more hands-on and a bit more granular than the rest of this book, but it's important to get the words right, as well as the ideas and the flow. There are a lot of moving parts in the process of writing, and they all need some grease if you want your manuscript to drive smoothly.

Chapter 6

Responding to Professional Feedback: The Most Important Writing You Will Ever Do

6.1 Preparing the response

Okay, so you wrote your manuscript and sent it off some time ago. You poured your heart and soul into it. And now the reviewers have responded... and they hate it. They're tearing it to pieces. This is awful and makes you want to curl up into a ball, right?

Wrong. If you've reached this point, congratulations! You're over the hump. Most manuscripts never get this far. They get submitted and then rejected by the editor long before they get sent out for review. If yours has made it to the tearing-to-pieces stage, you're already ahead of the game.

What's the point of peer review anyway? If your work can pass muster among people who do not care about you in the slightest — and who can ask probing questions and improve the quality of what you have — then you know that what you've achieved is objectively good. Is this easy? No. Will it make you feel warm and fuzzy? Also no. But if it were easy, everyone would do it.

Of all the writing you'll ever do, responding to reviewers is the most important. I once worked for a very intense boss, who would pick fights with everyone. She would take on colleagues, editors, random strangers, you name it. And yet, the only time she was scrupulously polite was when writing the response to reviewers. I learned a valuable lesson from that.

The important thing is to be professional. You should be polite and courteous, but you don't need to be obsequious. If you disagree with something the reviewers say, you can present your case... but you also need to present a better solution than you originally had and a better solution than they offered.

What you shouldn't do is argue with the reviewer, even if you think they're a moron who doesn't understand your genius. Even if you're right. They hold all the power, and you hold none, so arguing in this situation is a fool's errand. What you do need to do is address every point. Absolutely do not ignore something they raise.

Personally, when I peer-review papers, I'm fine if people disagree with me and present evidence. I'm more than fine if they take on board my wise observations. What really gets my goat is when they ignore my points. I work hard to help the authors turn their manuscripts into something of high quality, and I don't appreciate that hard work being ignored.

A killer response is 'We have done everything the reviewer requested.' (And of course, you have actually done those things!) With a response like that, how could they ever reject you? So try to make every change they ask for if you possibly can.

If you really, truly can't, then you do have a get-out-of-jail-free card, but don't overuse it. If the reviewer has asked for something truly ridiculous, the nuclear option is to say 'This is outside the scope of our manuscript.' But only save that for emergencies. Out of more than a hundred peer-reviewed papers that I've written, I think I've used that phrase three times.

WORKED EXAMPLE: *'This paper catalogs new research in population modelling. The intent is 'mapping the population modelling domain by examples of work.' Eighteen authors submitted brief synopses of their work as a follow on article to an earlier work by the Population Modelling Workgroup of IMAG. The article is useful in that it introduces a number of new authors and studies in population modelling and can therefore help build the community of scholars. Still, as an effort to map, it falls short of its potential. Traditional maps have organising principles — directionality (N,S,E,W) and distance — render them useful. Newer computer maps (e.g, Google Maps) add zoom in and out capabilities as well as different views (traditional, photographic, 360, etc.) The point is that the authors would be well served to spend a bit more time introducing the concept of population modelling and that concept's evolution and then, in the discussion, summarising the effects of the research of these new authors and research on the map. That would enhance the article by moving it to a summary of data points on research to a better map of the domain.'*

This was perhaps the toughest review I ever got. It was for a summary paper that collected together snapshots of disease modelling results by a large number of authors, in an attempt to map the field. The reviewer picked up on the word 'map' and essentially asked us to add in a magnification capability... in a research paper.

Huh? How the hell were we supposed to do that? I almost abandoned the manuscript at that point, because I couldn't begin to conceptualise what he wanted.

Now, if you read through the review carefully, you'll see that — technically — the reviewer isn't actually asking us to do that, instead asking us to spend more time introducing concepts and summarising the effects of the research. Those are all easy-peasy. But the magnification thing is clearly important, or else it wouldn't be mentioned in the review. Reading between the lines, we need to address this... somehow.

Inspired by this, I thought long and hard about how to achieve an equivalent of new technology in a summary paper. And I came up with something that worked: a table. But not just any table, a way to visualise the interacting layers of contributions, arrayed by both topic and method. See Figure 6.1. And so this allowed us to write a killer response to the review.

RESPONSE: *'We would like to thank the reviewer for providing us with a challenge... that we were extremely pleased to meet. Organising such a disparate group of topics into a coherent structure is not an easy task, nor is it a unique one. However, the reviewer's observation about the technological utility of maps inspired us to think somewhat laterally about this issue. As a result, we have taken three (perhaps four) approaches to answering this question:*

I. The examples are grouped by application topic in the manuscript itself. We have also provided a table at the end of the manuscript that gives a different structure comprising:

II. A different take on the application topics

III. A summary of the methods used

(IV. An indication of the overlapping nature of these topics.)

We are extremely pleased with the result and are indebted to the reviewer for the suggestion. In addition, we have added details of this reasoning to both the Introduction and the Discussion (in red).'

Unsurprisingly, the paper was accepted [Smith? *et al.* (2016)]. Well, wouldn't YOU accept it with this response?

Contributor	Managing disease spread	Resource planning and allocation	Predicting drug effects	Risk assessment	Ecosystem management	Testing theory	Epidemiology and public health	Summary of Methods
Robert Smith?	x		x				x	Ordinary and impulsive differential equations, Latin hypercube sampling, Monte Carlo simulations
Bruce Y. Lee	x						x	Agent-based models
Aristides Moustakas	x							Agent-based models
Andreas Zeigler				x				Random forests, support-vector machines
Mélanie Prague	x		x				x	Ordinary differential equations with nonlinear mixed effect models, control theory
Romualdo Santos		x			x			Differential equations, difference equations, Malthusian modelling
Matthias Chung						x		Robust and efficient point estimator methods for ordinary differential equations
Robin Gras					x	x		Agent-based models, fuzzy cognitive maps
Valery Forbes				x		x		Matrix population models, individual-based population models, dynamic energy budgets, mechanistic effect models
Sixten Borg		x	x				x	Finite mixtures of disease activity models, cost-effectiveness analysis
Tracy Comans		x					x	Discrete-event simulation of health services, cost-effectiveness analysis
Yifei Ma	x	x					x	Network models, database simulation, diffusion dynamics, multi-theory methodology
Nieko Punt			x					Pharmacokinetics/pharmacodynamics modelling, two-stage Bayesian parameter estimation
William Jusko			x			x		Pharmacokinetics/pharmacodynamics modelling, ordinary differential equations
Lucas Brotz				x	x			Fuzzy logic analysis of population dynamics to investigate trends
Ayaz Hyder	x			x		x	x	Agent-based models, microsimulation models, cost-effectiveness analysis, computational exposure science

Fig. 6.1. The nearest equivalent to zooming in on a paper

6.2   A template for success

An easy thing to do that many people miss is to simply highlight any changes due to the reviewer in colour in the manuscript itself. If there are two reviewers, highlight the changes due to one in red and the other in blue. If there are more reviewers, choose more colours, but make them distinct. This draws the eye and also focuses the reviewer on the changes you've just made, rather than forcing them to reread the entire manuscript (and running the risk of them finding further issues).

As mentioned above, a reviewer wants to make a positive impact. So you need to put those impacts in the manuscript itself. Too often I see people writing a perfectly decent rebuttal to a point the reviewer made... without actually putting any of that in the manuscript itself.

Here's an example response that you can use as a template.

WORKED EXAMPLE: *'Dear Dr. Yang,*

We thank the Editor and Reviewers for their time and consideration of our manuscript on RSV vaccination. We have done everything the reviewers requested. Here is a point-by-point response to the reviewers.

Reviewer 1

General This reviewer had four major comments and three minor comments.

Response: We have done everything this reviewer suggested. Changes due to this reviewer are in <u>blue</u>.

Comment (1) The authors describe current efforts at vaccine development as 'focused on the development of particle-based and subunit vaccines (p.2, 3/4 down)'. They do not mention, or consider, the continuing work on live attenuated vaccines or vectored vaccines both of which are supported by multiple large pharma companies.

Response: Good point. We have changed this sentence to: '...focused on the development of particle-based, subunit and vectored vaccines. Live-attenuated vaccines are also undergoing phase 1 trials.' This is based on the summary at http://www.path.org/vaccineresources/files/ RSV-snapshot-December2016.pdf *(Page 2)*

Comment (2) The authors consider (p.3, middle) 'a vaccine strategy for RSV where a fixed proportion of individuals entering the model are temporarily immune to infection. This reflects the situation where newborn children are vaccinated at birth.' They do not mention, or consider, that the main vaccine strategy now being pursued for the youngest infants is not direct immunization, but is instead maternal immunization. Vaccination occurring during the third trimester generates antibodies in the mother that are transferred transplacentally to the infant, resulting in higher antibody titers in the infant at birth. The thought is that the higher antibody titers should protect the infant for approximately two months longer. Pre-formed antibodies decay with time, and by 6 months maternal antibodies are no longer detectable in an infant.

Response: This is an excellent point, so we have changed the focus in this section to maternal vaccination and discussed this in some detail. Happily,

*by doing so, the results are unchanged from a mathematical perspective.
(Pages 3, 7, 16, 17)*

Comment (3) *Vaccination of infants as soon as they are born is seldom
successful for any pathogen because the infant's immune system is imma-
ture. It is not currently being contemplated for RSV. However, MedImmune
(owner of the prophylactic monoclonal antibody that is currently used for
'at risk' infants to protect them against RSV) has developed a more potent
RSV-neutralizing monoclonal antibody with increased stability that could be
given at birth to protect infants for their first 6 months. This approach
would avoid the uncertainties of individual maternal responses to RSV and
the problem of premature birth which could result in incomplete transfer
of the antibodies elicited by a maternal vaccine, depending on the time of
vaccination relative to birth.*

*In general, the thinking in the field is that there will be two vaccines
for RSV, one to protect infants during their first 6 months, and another
to protect them from 6 months on. I realize that there may be too many
variables for the authors to consider in one report, but they could choose
one of these strategies and model that. Maternal vaccination before birth
would seem to be the most important to study now since it is being
pursued aggressively by the NIH and two big pharma companies and several
smaller companies, and is being supported by the Bill and Melinda Gates
Foundation. The MedImmune stabilized monoclonal antibody approach
could be included as generally equivalent.*

Response: *This brings up a point that we realise was not clear: we are
actually considering both options. The nonimpulsive model considers pre-
infection vaccination only, while the impulsive model considers subsequent
vaccination. We have added emphasis in several places to make this clear.
(Pages 3, 7, 11, 16, 17)*

Comment (4) *But the protection of any of these approaches would cover
only the first 6 months of life. Thereafter, immunization of the child with
another vaccine would be needed to induce active immunity and a recall
response that would provide future, more rapid protection upon infection.
Right now, live attenuated or vectored (adenovirus) vaccines are the front
runners, but direct immunization with a subunit vaccine might eventually
be considered. A subunit vaccine has not been considered largely because of
the initial formalin-inactivated vaccine trial in the 1960's resulted in much*

more severe disease following the first community acquired infection in the vaccinees.

While modeling a 10-year protective vaccine and a lifetime-protective vaccine can be done, even infection with the wild-type virus does not provide 10-year protection, so it is difficult to see how a long-term protective vaccine could be generated. Nevertheless, it is a laudable goal.

Response: *This is a helpful observation. We have decided to change our focus away from long-lasting vaccines and instead mostly focus on short-term durations, as the reviewer suggests. We have mostly restricted ourselves to vaccines lasting six months (corresponding to $\omega = 2$) and have instead moved the focus to vaccine coverage via the proportion of individuals who are vaccinated (r). We re-ran all our simulations and have thus updated all figures. The results are actually stronger with this new focus, so we are grateful to the reviewer for raising this. (Pages 12, 13, 14)*

Comment (5) *p.1, author list. Why is Robert J. Smith followed by a '?' ?*

Response: *It is part of the author's name. See, for example:*
    http://mysite.science.uottawa.ca/rsmith43/MDRHIV.pdf

Comment (6) *p.18, l.10. vaccination-induced*

Response: *Fixed (Page 16)*

Comment (7) *p.18, l.20. outcome than coverage*

Response: *We agree, although this sentence has now been deleted, so it no longer applies.*

Reviewer 2

General *This reviewer noted that the research questions examined in our manuscript are extremely important and relevant and that we use an innovative approach to address the question of potential vaccine efficacy. This reviewer had five major comments.*

Response: *We have done everything this reviewer suggested. Changes due to this reviewer are in* red .

Comment (1) *The authors base the model on the assumption that infants will be given the vaccine at birth. While this is true for a few vaccines, most are not given at birth. Additionally, the most advanced vaccines in* development are not being targeted to infants. They are primarily targeting the elderly, and pregnant mothers to protect newborns. The authors need to address the fact that their assumption is very unlikely, or even false more than they have as the manuscript stands.

Response: *This is a good point that was also raised by Reviewer #1. See our response to Comments (2) and (3) above. Note in particular that we are actually considering both and have made that more clear. (Pages 3, 12, 16, 17, in blue.)*

Comment (2) *The authors conclude that vaccine duration would be more important than vaccine coverage. They recommend that vaccine be tested for duration before approval. The authors need to discuss how the practicality of studying long term immunity is very challenging, especially regarding the time frames they test. Obviously 70, or even 10 years would be impossible to test during a clinical trial before licensure.*

Response: *This point was also raised by Reviewer #1. We have changed the focus to short-term durations and re-run our simulations. See our response to Comment (4) above. (Pages 12, 13, 14, in blue.)*

Comment (3) *Vaccine duration is a somewhat vague term, especially since the correlates of immunity have not been fully defined for RSV, and natural infection does not necessarily confer protection from reinfection.*

Response: *This isn't as important now, although we will note that it is a well-defined term mathematically, even if that is an approximation to a more fuzzy concept in reality. We have added a definition. (Pages 3–4)*

Comment (4) *The authors should cite other, already licensed vaccines that are in use where duration is more important than vaccine coverage.*

Response: *We have changed the focus away from this, although we did find that this is true for both pertussis and HPV.*

Comment (5) *The endpoints of most RSV clinical trials are not sterilizing immunity, but a reduction in RSV-associated hospitalizations. The authors should consider incorporating this endpoint into their model or at least discuss this point.*

Response: *This is a good point that is worth mentioning. We have added a paragraph to the Discussion addressing this. (Page 17)*

In summary, we feel that these revisions have addressed all the points raised by the reviewers and hope that the manuscript is now acceptable.

Yours sincerely,

Alexandra Hogan, Geoffry Mercer and Robert Smith?

Important points to note are:

- Address the editor (they have the discretion to accept your manuscript without sending it back to the reviewers, although this happens only rarely).
- Indicate what you've changed and where to find it in the manuscript.
- Make sure the changes are actually in the manuscript!
- If you keep something intact, be clear about that... but it also probably means you need clarity anyway, so re-word it (even if you don't think it's necessary). See the response to Reviewer 1, Comment (3), for example.
- If the reviewer compliments your work, include it. See the initial response to Reviewer 2.
- If you can't change something, incorporate the concerns in the Discussion. See the response to Reviewer 2, Comment (5). But don't overdo this. Mostly you SHOULD be making changes at this stage! (We only added this to the Discussion because the reviewer gave us an out in the 'or at least discuss this point' comment.)
- Always be polite. Don't be snarky or combative.

Here is the reworked text from Pages 3–4, addressing Reviewer 1 Comments (2)–(3) and Reviewer 2 Comments (1) and (3).

WORKED EXAMPLE: 'Section 2: The nonimpulsive model

We first extend the SEIRS compartmental model for a single age cohort described by Weber et al. (49) to include a vaccine strategy for RSV where a fixed proportion of individuals entering the model are temporarily immune to infection. This reflects the situation where pregnant women are vaccinated in their third trimester, generating protective maternal antibodies that are transferred transplacentally to the unborn infant, conferring protection from RSV infection in the first few months of life. We assume that the leaving rate μ is unchanged across all classes and that there is no disease-specific death rate. We scale the entry and leaving rates so that the population is constant.

Let S represent susceptible, I represent infected and R represent recovered individuals, with V, I_V and R_V the corresponding compartments for *vaccinated individuals. The birth rate is μ, with a proportion p vaccinated, of whom ε successfully mount an immune response; the death rate is equal to the birth rate. The time-dependent transmissibility function is $\beta(t)$, with recovery ν and loss of immunity γ. The transmissibility of infected vaccinated individuals is described by $\beta_V(t)$, and the recovery and loss of immunity rates for vaccinated individuals are ν_V and γ_V respectively. Finally, the waning of the vaccine protectiveness is given by ω. Note that, although the definition of vaccine duration is not fully elucidated for RSV, mathematically it is well-defined as the period spent in the vaccinated classes before returning to the associated unvaccinated classes. This definition is based on an exponentially distributed time spent in the vaccination classes, and hence the duration corresponds to $\frac{1}{\omega}$ years.'*

Finally, do not be afraid to make big, substantial changes. This is the time when your manuscript is in flux, with the added benefit that you have some considerable distance from it.

My first post-Ph.D. paper had the reviewer asking me to change the mathematical model. I really resisted this at first, because it would mean reworking the entire manuscript, re-doing all the analysis and the numerical simulations. But I did it anyway, and consequently the changes had to pass through the analysis and necessitated new figures. And you know what? The results were basically consistent with the old model. This gave me a huge boost of confidence in my work, because of course the results shouldn't be so sensitive to changes in the model. And when I later presented this work at conferences and got asked questions, I really knew all the ins and outs of how the model worked and why every piece was there, thanks to the changes I made because of this reviewer.

BAD EXAMPLE : *'Review Comment 5: Properly explain figures 12–14. What effects occurs on different controls by varying the values of weight constants. The upper bounds of controls change. What does it biologically mean? How we explain it for general readership?*

Author's response: I don't see why the respected reviewer is asking for the biological interpretation of the upper bound of a control. The upper bound is just the maximum amount of the control that can be applied, there is no other interpretation more than that. These figures are obtained by checking the relationships among the three controls. Although, we briefly discussed the relationships among the controls when varying the weight constant associated to the controls variables.'

This was the original draft that my co-authors wrote in response to a review. Although they use the words 'respected reviewer', you can pretty much tell from the tone that they are displaying anything but respect. The reviewer will absolutely sense this. You do not want to annoy the reviewer. If nothing else, they're simply doing their job. At best, they're saving you from yourself.

Also, it's a reasonable request. From a distance, that's probably obvious, but I understand the author's frustration when being questioned about things that seem either obvious or irrelevant. But... you have to do them anyway. Take the revisions like a grown-up.

(RE)WORKED EXAMPLE: *'We have completely rewritten this section in order to address these points, which are very valid. (Note that we have removed a figure, so these are now numbered 11–13.) We have explained the effect for a general readership, with biological meaning. To improve the flow, we have also given these results their own subsection. (Pages 20–21)'*

There. That's more like it.

BAD EXAMPLE : *'Comment 9: How can we say that backward bifurcation occurs at R0 = 1. Please explain it mathematically.*

Author's response: The phenomenon of backward bifurcation occurrence in disease transmission models, where a stable endemic equilibrium co-exists with a stable disease-free equilibrium when the associated reproduction number is less than unity, has been observed in a number of disease transmission models. The epidemiological consequence of backward bifurcation is that the classical requirement of the reproduction number being less than unity becomes only a necessary, but not sufficient, for disease elimination (hence, the presence of this phenomenon in the transmission dynamics of a disease makes its effective control in the community difficult). In order to have mathematical results for the backward bifurcation in a mathematical model proposed in our paper, we can have the results in supplementary file in section 'Analysis of backward bifurcation' using the results described in Ref [6]. There is another way that can be used to check whether there exists a backward bifurcation for a biological model, by obtaining the solution at the endemic equilibrium, if the solutions occur in the form of quadratic equation, then there may be a chance of backward bifurcation with specific values of the parameters, but some models give a cubic and high order polynomials then, a study is required to obtain the unique endemic equilibrium, in such a case if more than once endemic equilibrium exists then it may not have. The case to our problem, and the endemic equilibrium computed in Supplementary file section 'Analysis of endemic equilibrium' which clearly demonstrates the non-existence of backward bifurcation for the given model'

Okay, so this is a total rookie mistake. The reviewer isn't asking what a backward bifurcation is, they're asking the authors to explain mathematically *in the manuscript itself*. This response wastes a lot of words lecturing the reviewer about something they already know. And the final point seems to say... that there isn't, in fact, a backward bifurcation.

(RE)WORKED EXAMPLE: *'This has also been removed, as the previous version was incorrect. With global stability of the disease-free equilibrium, no backward bifurcation can occur.'*

BAD EXAMPLE : *'Comment (2): L 269: Is this project aiming to control or to eliminate?*

Response: This project aims for both controlling and eliminating soil-transmitted helminthiasis in Kenya (4).'

Okay sure... except that this answer only went to the reviewer. The changes need to appear in the manuscript itself.

(RE)WORKED EXAMPLE: *'This project aims for both control and elimination (4). We have added a note to this effect. (Page 14, lines 311–313)'*

Much neater.

BAD EXAMPLE : *'Comment (7): L 3: Can you make any comment regarding the cost of using an additional round of weaker drugs vs cost of using stronger drugs and fewer rounds?*

Response: We would like to apologize for unable to tell you which one is more cost efficient (either using an additional round of weaker drugs or using stronger drugs and fewer rounds). This is because as we do not have any data/information about this question at the moment but we could provide you some information regarding cost of STH [soil-transmitted helminths] treatment as follows: By treating children in school, a higher treatment coverage can be achieved and much cheaper in cost compared to treating adults. Nevertheless, treating children alone does not always reduce the transmission significantly, this is especially the case for hookworm as the adults are the major drivers of transmission (3). Although by scaling up the treatment to the whole community will lead to significant reduction in transmission, the main issue is the cost of treatment, which is depends on demography, coverage level, population that is going to be treated, frequency of treatment, and for how long. If the interruption of transmission does not happen by treating children alone, treatment has to be continued forever unless WASH [water, sanitation and hygiene] programmes and education have potential to change this condition (3, 8).'

Once again, there's no need to lecture the reviewers, who are more than likely experts in the field anyway. The Response to Reviewers exists to be in service to the manuscript. Even if there's no data, the manuscript still needs to be tweaked.

(RE)WORKED EXAMPLE: *'Unfortunately, we do not have any data about this question at the moment; however, we have provided some information regarding cost of STH treatment in children vs adults (Page 2, lines 48–57)'*

This streamlining wastes a lot less of everybody's time, and it means the reviewer has contributed to making the manuscript better. Everybody wins!

6.3 Further polishing

There's one other thing that the review process can do that's good to pay attention to: it can cause you to question everything. More than once, a reviewer has raised what looks like a minor point, one that could be fixed with a small change... but I've realised that it's actually a major issue that they've missed. It would be easy to just make the small change they're asking for and resubmit, but I'm going to recommend that you don't do that, for several reasons.

First is that your end goal here isn't just Yet Another Paper. That's nice, of course, but you actually want work that's good and readable and which can hopefully make a difference in the world. If you knowingly publish erroneous work, you're doing yourself and your academic field a disservice.

If you find out later that you made a mistake, own up to it in a future article or an erratum. Take it like a grown-up, and be clear about what happened so that future researchers can untangle things.

The second reason to rework things is that you're not actually guaranteed an acceptance with the minor change. Restructuring whole sections because of a question the reviewer raised will impress them no end and is sometimes what they actually want, even if they aren't able to articulate it.

And the third reason is that this is (almost) your last chance to change your manuscript. You can do a proofread after acceptance, but that's just a polish and is partly to catch any errors introduced during the publication process. You won't be able to make major changes after this, so seize the moment. You'll regret it if you don't.

In fiction writing, there's a saying called 'Kill your darlings.' What this means is that you shouldn't be afraid to delete your favourite parts of your story if they're not in service of the plot. Or, as a famous writer I talked to once said, 'I knew I was a real writer when I cut my favourite scene from the book.'

The same applies here. You spent a lot of time and effort on this manuscript, so you probably have bits that are meaningful and important. But if they need to go because they're not working, then cut them or rewrite. It's tough to do, no question... but you'll be a real writer when you can do that. (I once had to delete my favourite sentence from a manuscript and boy did that suck. It was the right call for the paper, but I still hated doing it.)

Remember that the goal is and always has been this: make the article the best version of itself it can possibly be. Do whatever you have to in order to reach this goal.

WORKED EXAMPLE: *Remember how the major selling point of the RSV manuscript was those unexpected infection spikes shown in Figure 1.7? You know, the ones that found their way into the title. Yeah. About that.*

It's not that there aren't any, it's just that the various changes made throughout the revision process have had a knock-on effect, and the infection spikes now look quite different. Figure 1.7 has now become Figure 6.2, complete with a more detailed caption.

Part of the issue here was that the earlier version didn't look so clear, but it also didn't match the form of the other figures. And I think it wasn't terribly obvious that the top figure in Figure 1.7 was actually the case of no vaccination. I think the revised figure tells the story a lot more visually.

There are a couple of notes embedded in this figure caption. The first is just to highlight the different timescales. Looking at it now, I possibly

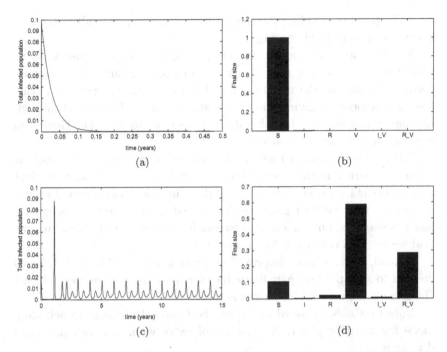

Fig. 6.2. Extreme parameters show that perfect vaccination can induce unexpected infection spikes. a. With no vaccine ($r = 0$), the result is that the infection clears and the entire population remains susceptible. (Note that the timescale is given for only 0.5 years to show the decline but was run for 15 years.) b. The final size of each compartment in the case of no vaccine after 15 years. c. When an imperfect vaccine is given to the entire population ($r = 1$), the result is a series of disease spikes in the vaccinated population. Note that the transmission rate is not oscillating in this example. d. The final size of each compartment in the case of full vaccination after 15 years. Vaccination thus destabilises the DFE.

should have done something similar with the slightly different vertical scales on the right as well. It's not terrible, but it's a bit annoying that they don't match, and there's no real reason for that. I think that if the top right graph had been capped at 1 so that it ran all the way to the ceiling on the box, the story would be even stronger.

The other note about the non-oscillating transmission rate is because somebody asked me that question during a presentation. I'd set up a whole thing about seasonal oscillations, and it hadn't really paid off... but there were now different oscillations happening, so I wanted to disambiguate the two. (With more time, I would have gone back and set this up more carefully, perhaps avoiding the oscillating transmission rate altogether, as it never ended up playing a big role.)

You have another opportunity at this stage: you might just want to make the change for the sake of clarity. Considerable time will often pass between submitting the manuscript and dealing with the reviews, so this has the advantage of giving you more distance from it. If you read through the manuscript and find bits that don't make sense to you, then there's no way they'll make sense to anyone else.

With some distance, I realised the timescales were ridiculous. Look at Figure 1.3 earlier in this book. You can hardly see what's going on. And nobody needs 150 years of the same old thing in order to understand what's happening. Oh, and the legends needed to be deleted. (They were originally just placeholders, but sometimes placeholders can start to look so familiar that we come to take them for granted.)

Instead, we have much-improved graphics in Figures 6.3 and 6.4. I also decided to split the four-part figure into a pair of two-part figures, which I think made life a lot easier for the reader.

Note that nobody asked me to fix the figures. But doing so definitely made for a better paper. A little bit of extra work now can pay huge dividends later.

WORKED EXAMPLE: Here was the response to the second round of reviews. Reviewer #2 was satisfied with our changes, but Reviewer#1 had more comments. You'll notice that things we didn't need to change have a clarification added anyway.

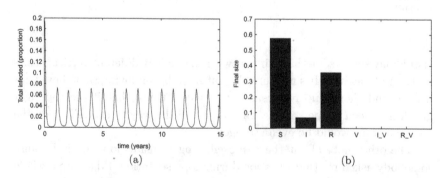

Fig. 6.3. Without vaccination, the disease infects up to 7% of the population. a. The total infected population, including vaccinated individuals. b. The final size in each population.

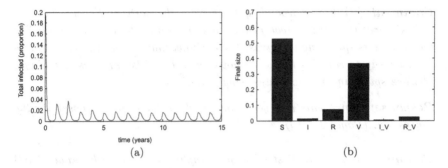

Fig. 6.4. 50% coverage with a vaccine that reduced transmissibility by half and waned after two years resulted in a substantial reduction in the disease compared to no vaccination. a. The total infected population, including vaccinated individuals. b. The final size in each population.

'Dear Dr. Yang,

We thank the editor and the reviewer for their time and consideration of our manuscript on RSV vaccination. We have done everything the reviewer requested. Here is a point-by-point response to the reviewer.

Reviewer 1

General This reviewer notes that the modified manuscript is improved and had five minor comments.

Response: We have done everything this reviewer suggested. Changes due to this reviewer are in blue.

Comment (1) Why is Smith followed by a '?' in the author list?

Response: It is part of my name. See, for example:

http://mysite.science.uottawa.ca/rsmith43/MDRHIV.pdf

Comment (2) p.1, next to the last line mentions 33.8 million episodes of RSV. Does this mean infections, or RSV disease that is severe enough to require medical attention? There is a 10-fold difference between these two possibilities.

Response: The former. We admit that the cited paper is less than clear on this in its Abstract, but it is definitely infections. We've added a clarification. (Page 1)

Comment (3) *Fig. 8 legend, 4 lines from the bottom 'Vaccine-induced spikes'. Does the vaccine really cause this amount of disease? It sounds like the vaccine is spreading the virus. I don't think that you mean that. Is there a better way to say this that does not lead to this false impression? Maybe 'disease spikes in the vaccinated population.'*

Response: *Done. This is a good point and we thank the reviewer for the suggestion. (Pages 15, 16)*

Comment (4) *p.17, first blue paragraph: 'some (poorly understood) existing antibodies' What do you mean by poorly understood? These are well understood antibodies. They come from the mother, cross the placenta by a receptor-mediated, specific transport mechanism for IgG only, and they represent the mother's serum antibodies during the third trimester.*

Response: *What we wanted to say is that the role of maternal antibodies in helping newborn infants develop protective immunity to RSV is not well understood (as in, the level of antibodies required, and how protective they are). However, the reviewer is correct in that maternal antibodies themselves are relatively well understood. Since this was misleading, we have decided to simply delete the words '(poorly understood)' to avoid any confusion. (Page 17)*

Comment (5) *The results in Fig. 6 suggest that a 75% vaccination rate would lead to eradication. Why not say that in the Abstract? But 50% vaccination could lead to spikes in virus infection (Fig. 7). Say that next.*

Response: *Done. We thank the reviewer for the suggestion, as it improves the Abstract significantly. (Page 1)*

In summary, we feel that these revisions have addressed all the points raised by the reviewers and hope that the manuscript is now acceptable.
   Yours sincerely,

   Alexandra Hogan, Geoffry Mercer and Robert Smith?'

I don't know why that reviewer was so hung up on the question mark in my name, but it made it through two rounds of reviews. Incidentally, if you go to that link, you'll discover my first-author paper in the journal *Science*, which had no problem including the question mark. I figured that if it was good enough for *Science*...

6.4 Acceptance

Okay, so the best outcome has happened and your article has been accepted. Hooray!

What happens at this point is that the paper will go to production, where it may be edited. You'll receive the galley proofs (laid out the way the article will be) and a short window of opportunity to make any last corrections or changes.

Absolutely take this opportunity! I was appalled when I discovered my ex-girlfriend didn't even look at the proofs of her articles, as she only cared about acceptance. (I offered to do them for her, and boy were there critical things in there.) The journal may make changes that you're not okay with or that entirely change the meaning. You need to be vigilant.

You're usually only given 48 hours to make these changes. Depending on the journal, you might be able to ask for an extension, but you might not. I once got a set just before heading out to Australia, so I had to spend my plane ride proofreading and, as it turned out, making major changes, as I discovered a huge error and had until the plane landed to fix it.

If English isn't your strong point, get someone on your team who's good at it to do this proofread, as it's critical. Most egregious errors happen at the last minute because everyone is rushed and just wants the process to be over. Don't compromise on quality at this stage.

Once this is done, you'll have no further chance to make any changes, even if you discover something later. If it's particularly terrible, you can publish an erratum or, in the worst-case scenario, retract the article. But hopefully you've done enough prep work through this process that this shouldn't be necessary.

If you want to see how the final version came out, after all this work, check it out here:

https://mysite.science.uottawa.ca/rsmith43/RSV.pdf

So there you have it. The entire publishing process, from beginning to end. Only... there's one more chapter left. So what comes next?

Chapter 7

What's Next? Talking the Talk

The sad reality is that most people don't have time to read. You put a huge amount of effort into writing a paper... and the response is often just crickets. That doesn't mean you shouldn't put the effort in, because the crucial people who do read it will notice, but it can be disheartening when you send it out into the wild and there isn't much response.

However, they do have time to listen — at conferences, in seminars, online etc. So the last thing to do is to turn your paper into a 20-minute academic talk. Ideally, you want to do this before the final version, because the process of converting a paper into a talk is incredibly effective at finding outstanding issues or gaps in logic, which will allow you to iron out any remaining weak spots.

Make sure you actually present this to your colleagues. Universities are always looking for speakers for seminars, so you shouldn't have trouble finding a slot. It's important to get their constructive feedback and questions, both for justifying your own logical arguments and also for finding any holes in the research or the presentation.

This book is about writing, not talk-giving (that'll be a future book!), but your talks should follow the same principles outlined here: have academic depth, but don't be boring — and for goodness' sake have a narrative!

People are more likely to be exposed to your talk than to read your paper, so don't skimp on it. In particular, *never* prepare your talk while at the conference itself. Not only is it rude (you should pay attention to other talks as you would want people to pay attention to yours) but you won't know the timing or if some constructions work. And you need to practice, practice, practice. Ideally, you want at least three practices, on three separate days: once for saying the words, once for timing and once for a polish.

The academic talk showcases your professionalism, which is built out of the research, and your reputation will matter in the future. Think of it as the big-budget movie version of your paper. You want to educate, enlighten and entertain.

The talk is the major way you get other people interested in your work. It needs to be deep, dramatic and digestible. Use lots of visuals and pictures (talks and papers are different things), and don't 'read out' your slides. Do not go over time (it's, again, rude, and your advance practice should take care of this). Be sure to leave time for questions.

Big questions may come out of your presentation. Smart people who know nothing about your work may ask things like 'Have you considered...?' or 'What about...?' and these are *incredibly* valuable questions. Even if you don't have good answers in the moment, think about them afterwards. These may be the basis for your next paper. There's always more exciting research waiting!

Chapter 8

Summary: Don't Be Dull

Okay, so what did we learn that we didn't know?

- Research starts with an idea. It doesn't even have to be your idea, but you need one from somewhere.
- You find the paper through editing and refining. Never publish a first draft. Conversely, don't get hung up on your initial draft. Just write the thing, and worry later about how to make it work. You can't edit a blank page!
- Your guiding principle is 'What did we learn that we didn't know?' If you can't answer this, don't publish.
- The Abstract and the Introduction need to be tight. Edit, edit, edit.
- The figures need to tell the story. The rest of the paper adds to this for the curious reader, but the figures need to tell the story on their own.
- The Discussion is for implications. It's not an extended summary of what we just read.
- The paragraph is your main unit of writing, but every sentence counts. A single sentence that lacks logical cohesion can badly damage your paper.
- Use the reviews to help you find the narrative. Feedback is incredibly helpful, so treat it as the valuable resource that it is.

Ultimately, you're telling a story. It needs to have a beginning, a middle and an end. Make it a good one!

Answers to Homework Questions

Technically, there's no single 'right' answer. But Anthony suggested having my own rewrites for the three homework examples on Page 81 so you an compare your version with mine.

BAD EXAMPLE : *'The quest for a world without leptospirosis is achievable if the control and the eradication of leptospirosis is taken serious owing to it strong biological and socio economical importance.'*

(RE)WORKED EXAMPLE: *'Given the strong biological and socioeconomic importance of leptospirosis, its eradication is possible if the disease is taken seriously.'*

Here's Anthony's version, along with own grading of his attempt and comment:

> 'OK, let's see what I've learned. "A world without leptospirosis is one that can be reached and should be given its strong biological and socio-economic impact, but in order to achieve this, its eradication must be taken seriously." (Not sure. 8/10. Struggled with getting "biological and socio-economic importance" in and keeping it as a single sentence. I also may have changed the emphasis by adding a modal verb.)

If you struggled with this, as Anthony did, don't panic. It's a tough one given the two conditionals ('if' and 'owing'/'given'). An alternative would be to break it into two sentences or to drop one of the conditionals.

BAD EXAMPLE : *'The Tonado plot and the sensitivity analysis of our model revealed factors that mostly influence our model and in fact the real life as leptospirosis is concerned.'*

(RE)WORKED EXAMPLE: *'The Tornado plot and sensitivity analysis revealed key factors that influence the model, which help us understand the spread of leptospirosis.'*

Anthony's version:

'(Given that I don't understand what it's trying to say, as the model appears to be influencing the model, which I just don't get...) "Both the Tornado plot and our sensitivity analysis of the factors that affected the model clearly reveal those same factors which influence the spread of leptospirosis in real life." (6/10, but mostly because I don't understand it. I still think it's better; I just don't necessarily think it says what was intended, because I don't understand what that was. I'm also not sure it's about spread — it could be, for example, transmissibility, so I've made a potentially inaccurate assumption there.)'

He wasn't far off, but he's absolutely right that the model influencing the model was a stone around the neck of this sentence.

BAD EXAMPLE : *'Among all these parameters, our results strongly point vaccines efficacy as the most important parameter that can reduce the transmission of leptospirosis as any little increase in the vaccine efficacy will greatly reduce the basic reproduction number of our model and by implication reduction in the transmission process of leptospirosis.'*

Here's my first draft:

(RE)WORKED EXAMPLE: *'The most important parameter is vaccine efficacy, which has a significant influence on the reproduction number, which in turn will reduce transmission.'*

Here's Anthony's take:

'"The results of this study strongly suggest that vaccine efficiency is the most important factor in reducing the transmission process of leptospirosis. Moreover, it also shows that even a small increase in said efficiency has a marked effect on reducing the basic reproduction number." (9/10, I think, probably because I understood this one!)'

Actually, I think I like Anthony's version a bit better than mine, because I reduced the 'any little increase' part to the word 'significant', whereas Anthony actually ran with it. (Technical note: it's vaccine 'efficacy', not 'efficiency'. But not bad for a music teacher!) Here's my second version.

(RE)REWORKED EXAMPLE: *'Vaccine efficacy is the most important parameter in reducing the spread of leptospirosis. Small increases in vaccine efficacy can greatly reduce the reproducton number, hence limiting the transmission of the disease.'*

There, doesn't that feel a lot cleaner?

Bibliography

Aggarwala, B. and Smith?, R. (2009). Do frequent doses of HAART block the HIV blips? *Far East Journal of Mathematical Sciences* **32**, 231–251.

Coffeng, L. E., Truscott, J. E., Farrell, S. H., Turner, H. C., Sarkar, R., Kang, G., de Vlas, S., and Anderson, R. (2017). Comparison and validation of two mathematical models for the impact of mass drug administration on Ascaris lumbricoides and hookworm infection, *Epidemics* **18**, 38–47.

Crawley, G. M. (2015). *The Grant Writer's Handbook: How to Write a Research Proposal and Succeed.* World Scientific: Singapore.

Diskin, S. (2018). *The 21st Century Guide to Writing Articles in the Biomedical Sciences.* World Scientific: Singapore.

Fakhry, N., Naji, R., Smith?, S., and Haque, M. (2022). Prey fear of a specialist predator in a tri-trophic food web can eliminate the superpredator, *Frontiers in Applied Mathematics and Statistics: Mathematical Biology* **8**, 963991.

Guy, G. (2018). Saks vs. Macys: $(r-1)$ marches on in New York City department stores, *University of Pennsylvania Working Papers in Linguistics* **24**, 7.

Glasman-Deal, H. (2020). *Science Research writing for Non-native Speakers of English London.* Imperial College Press: London.

Jain, G., Yadav, G., Prakash, D., Shukla, A., and Tiwari, R. (2019). MVO-based path planning scheme with coordination of UAVs in 3-D environment, *Journal of Computational Science* **37**, 101016.

Joshi, H. (2018). *Tips and Tools: A Guide to Creative Case Writing.* World Scientific: Singapore.

Khan, M., Ahmed, L., Mandal, P., Smith? R., and Haque, M. (2020). Modelling the dynamics of Pine Wilt disease with asymptomatic carriers and optimal control, *Scientific Reports* **10**, 11412.

Lebrun, J.-L. (2007). *Scientific Writing: A Reader and Writer's Guide.* World Scientific: Singapore.

Lebrun, J.-L. and Lebrun, J. (2017). *The Grant Writing and Crowdfunding Guide for Young Investigators in Science.* World Scientific: Singapore.

Nather, A. (2015). *Planning Your Research and How to Write It.* World Scientific: Singapore.

Smith?, R., Lee, B. Y., Moustakas, A., *et al.* (2016). Population modelling by examples ii, in *Proceedings of the Summer Computer Simulation Conference*, pp. 1–8.

Smith? R., Li, J., Gordon, R., and Heffernan, J. (2009). Can we spend our way out of the AIDS epidemic? A world halting AIDS model, *BMC Public Health* **9**, S15.

Smith?, R. J., Hogan, A. B., and Mercer, G. N. (2017). Unexpected infection spikes in a model of respiratory syncytial virus vaccination, *Vaccines* **5**, 12.

Weber, A., Weber, M., and Milligan, P. (2001). Modeling epidemics caused by respiratory syncytial virus (RSV), *Mathematical Biosciences* **172**, 95–113.

Wu, Y., Gao, Y., Zhu, B., Zhou, H., Shi, Z., Wang, J., Wang, H., and Shao, Z. (2014). Antitoxins for diphtheria and tetanus decline more slowly after vaccination with DTwP than with DTaP: a study in a Chinese population, *Vaccine* **32**, 2570–2573.

Index

Printed in the United States
by Baker & Taylor Publisher Services